INDUSTRIAL ART IN ENGLAND

AN ENQUIRY INTO INDUSTRIAL ART IN ENGLAND

by

NIKOLAUS PEVSNER

CAMBRIDGE
AT THE UNIVERSITY PRESS
1937

CAMBRIDGE UNIVERSITY PRESS
Cambridge, New York, Melbourne, Madrid, Cape Town, Singapore,
São Paulo, Delhi, Dubai, Tokyo, Mexico City

Cambridge University Press
The Edinburgh Building, Cambridge CB2 8RU, UK

Published in the United States of America by Cambridge University Press, New York

www.cambridge.org
Information on this title: www.cambridge.org/9780521170659

First published 1937
First paperback edition 2010

A catalogue record for this publication is available from the British Library

ISBN 978-0-521-17065-9 Paperback

FÜR

CAROLA

CONTENTS

ILLUSTRATIONS

ACKNOWLEDGMENTS

In preparing and writing this book, I have received great kindness and valuable assistance from many sides. In the first place, I wish to thank Professor P. Sargant Florence, who proposed this enquiry to me, made me his research assistant in the Department of Commerce of the University of Birmingham, followed the development of my work with untiring interest and read over my proofs. The endowment of the appointment was due to the Academic Assistance Council in London and some friends of the University of Birmingham. It is with sincere gratitude that I take this opportunity of recording their generosity. They have not only enabled me to start on this research but also to become acquainted with a country which at the time was still strange to me. Miss Francesca M. Wilson, Mr Geoffrey Baker and also Mrs E. E. Duncan Jones, Mr Austin Duncan Jones and Mrs Constance Bachtin helped to improve the foreigner's dry and insipid style. My mother, Frau Annie Pevsner, at Leipzig, took the trouble of copying the whole of the text, in order to make the manuscript printable.

It would not have been possible for me to collect the data which the following chapters contain, had I not enjoyed the active support of the Design and Industries Association through the London Secretary at the time, Miss M. E. Pheysey, B.A., and Mr E. O. Ransome, then Honorary Birmingham Secretary, and of the Board of Trade Council for Art and Industry through Mr Frank Pick, the Chairman, and Major A. Longden and Mr G. L. Watkinson, joint Secretaries. I am unable to express my gratitude to all those who have proved helpful in individual cases, partly because this would mean publishing the names of firms that wished to remain anonymous. But I gladly take this opportunity of thanking Mr R. D. Best, Mr Noel Carrington, Mr T. A. Fennemore, Mr Edward Hald, Mr H. H. Holden, Mr C. Huxtable, Mr P. H. Jowett, Mr Keith Murray, the late Mr H. H. Peach, Mr Gordon Russell, Mr W. L. Stephenson, Mr A. L. Wallbank, Mr J. Waterer, Mr Hugh Weeks, Mr J. Wedgwood and Mr R. Wyborn. I feel also greatly indebted to the directors of those firms who kindly allowed me to illustrate their products, and to Mr R. de Cronin Hastings, Mr R. J. Richards and Mr Geoffrey Baker, who agreed to my using in this book articles published in the *Architectural Review*, 1936, and in *Trend in Design*, 1936. The greater part of the enquiry recorded in the book was completed before the end of 1936. What has been added since is indicated in the Note on the next page.

NIKOLAUS PEVSNER

London
March 1937

NOTE

In order to make this book up-to-date wherever it was possible I have thought it advisable to compile a few "stop-press" notes concerning developments of 1937. They appear as addenda on pp. 232 ff., and an asterisk (*) in the text is meant to refer readers to these notes.

N. P.

March 1937

INTRODUCTION

"What business have we with art at all, unless all can share it?"

William Morris, 1883

INTRODUCTION

The aim of this book is to give a survey of the conditions and the artistic value of design in English industries. To discover the conditions I had to go round and interview manufacturers, managers of department stores and shops, principals and teachers of art schools, commercial designers, artists and architects, and to collect information from them. To discuss the aesthetic value of English products involves criticism of present methods and suggestions as to practicable improvements.

Since it was Birmingham University that enabled me to undertake this research, Birmingham formed the natural centre of my investigations. I tried as far as possible to limit them to Midland firms and Midland institutions, that is to an area of 50 miles round Birmingham. However, such a restriction proved sometimes impossible and more often inadvisable. Several industries which are of outstanding importance in the production and use of modern designs are hardly represented in the Midlands. In other cases, Birmingham firms could give but a one-sided view of a trade or, for whatever reason, declined to give me evidence. Moreover occasionally it appeared essential to check certain results by comparing them with information which I could obtain only from particular firms in other parts of Britain. On the other hand, several of the most prominent industries in the Midlands had to be left out completely, because they are not concerned with artistic design. This applies to the heavy engineering, the screw, nail, nut, bolt, cycle, lamp, tyre industries, etc.

I need not say that I never tried to get in touch with all the firms or even all the important firms in one trade. It proved sufficient to visit a selection of characteristic works in each industry, covering as far as possible all grades between high-price individual work and cheap mass production.

From the beginning I wish to emphasize that I did not want to confine my enquiries to well-designed goods only. It seems to me a flaw in some of the publications on art in industry which have been brought out during recent years that, by dealing with good design only, they are compelled to leave out

90 per cent. of the national production. But to plead for better design in manufactured wares will remain a futile task, unless the commercial aspects of the problem, unpleasant as they may sometimes be, are carefully taken into account. Such considerations induced me to listen as attentively to the manufacturer's point of view as to the arguments of studio artists, craftsmen, and art school experts.

In some trades I saw only two or three firms, in some ten or more, according to what seemed necessary and proved feasible. But though cheap and expensive, well-designed and badly designed, hand-made or machine-made articles passed before my eyes, and had to be considered and judged, I admit that my main interest was with design in mass production. For here, I think, is another point which has not been sufficiently considered in many of the recent publications on industrial art and (to anticipate) in many recent exhibitions also. As long as it is only or mainly the expensive product which is improved and discussed, not much is accomplished. Three-quarters of the total income in England in 1928 was divided amongst those earning less than £250 a year. It is primarily of these people that we have to think, and not of the one-hundred-and-fortieth of the population who earn more than £2000 a year.[1]

I started my research early in 1934, carried on with it continuously until the summer of 1935, and have since then completed my material, kept it up-to-date and, while preparing another book on the development of architecture and applied art between 1890 and 1914,[2] tried to add some more information as to the historical premisses of the present situation. During these two years I visited 149 manufacturers, 15 department stores, multiple stores and shops, 14 art schools, and 17 designers, artists and architects.

With manufacturers, as a rule, the following questions were discussed: General aspects of the trade, usual conditions of design in the trade, history of

[1] Cp. C. Clark, *The National Income*, 1924–31, p. 76, Macmillan, 1932. In Germany, where more exact statistics are available, the corresponding figures for 1932 show that almost 90 per cent. earn no more than 250 RM. a month, and about 65 per cent. up to 125 RM. a month ("Lohnsteuer und Einkommensteuer-Statistik", *Wochenberichte des Instituts für Konjunkturforschung*, 1932, Heft 7).

[2] *Pioneers of the Modern Movement* (from William Morris to Walter Gropius), Faber and Faber, 1936.

the firm, size of the firm, existence of a special drawing office or design department, existence of a trained head designer, participation of other persons, such as the manufacturer himself or managers, salesmen, foremen or workers in the designing, antecedents and payment of designers, relation to art schools, general importance of local art schools for the trade in question and for the particular firm, experience with commercial artists, free-lance artists and architects, and experience of the popular taste in designs, so far as shown by the latest sales figures.

In department stores, multiple stores and shops the one question was the appeal of the various goods to the buying public. But since this to such a large extent depends on the way in which they are presented, and since considerable sections of the public may not take to certain types of products simply because the shop does not give them a fair chance to see and compare them, it will never be sufficient to consult the manufacturer and the shopkeeper only; the public itself should be consulted. Much too little has been done so far to induce consumers to express their opinions of the goods which they are offered. In connection with this book only one such experiment could be carried out, and that of a very limited import.

In the summer of 1934 the Design and Industries Association together with the Birmingham City Art Gallery arranged an Exhibition of Midland Industrial Art. I suggested compiling a questionnaire which visitors were asked to fill in.[1] The organizers of the exhibition agreed, and a text was drawn up which contained the following questions: Sex? Age? Profession? Which objects in each department of the exhibition (electrical goods, hardware, furniture, glass, pottery, jewellery and silverware, textiles, other goods) do you like most? Which least? As I shall frequently have to refer to this questionnaire, I want from the beginning to stress its evident shortcomings. Only 92 answers were given, 30 female, 59 male, which is not a high percentage when set against the total number of visitors. Of these replies 11 came from visitors under 20 years of age, 41 from those between 21 and 30, 8 from those between 31 and 40, 10 from those between 41 and 50, 10 from those over 50. Twelve preferred not to disclose their ages. This proportion between the different age groups means that the answers may

[1] Cf. N. Pevsner, "A Questionnaire on Industrial Art", *Design for To-day*, vol. iii, p. 145, 1935.

well give an idea of the tastes of the younger generation, but not of the tendencies prevailing among those who in England represent the main purchasing power. And, furthermore, it was not even the younger generation as a whole that expressed its predilections and dislikes in the questionnaire. There was a remarkable class limitation in the answers. While only 6 were given by industrial workers, 4 by engineers (whatever that may be), 12 by clerks, accountants, secretaries, salesmen, etc., 4 by manufacturers and industrial managers, 4 by doctors, lawyers and other professional men, 6 by housewives, there were as many as 17 answers from students and teachers, and 18 from artists, commercial artists, designers, etc. Thus the answers indicate mainly the taste of young people with a distinct interest in educational and artistic matters. But even so, the results of this first experimental questionnaire may be useful; at any rate in conjunction with other arguments. It may be well, in this context, to draw attention to the fact that in a questionnaire on industrial products in the United States the answers from 25 students were up to 70 per cent. identical with those from 390 people of mixed social status.[1] This shows that perhaps a few witnesses belonging to one special class may be a fairer substitute for the public taste in general than might be expected.

So much for answers from the public itself. In discussing the problems in question with artists, architects and commercial artists I had to find out about their experience in designing for industry and in particular their experience of the manufacturer's attitude in cases where they had come into touch with him. Moreover I had to discover whether they were willing and apt to start working for commercial production, and to try and bring them into contact with firms interested in modern design.

Visits to art schools were made in order to verify the extent to which they were co-operating with industry. Points discussed were: general support from manufacturers, evening classes and afternoon classes for apprentices and journeymen, full-time junior classes for special industries, full-time senior courses for industrial design, attempts at any independent commercial production in the schools and the sale of their products, connection between training in design, handicraft and the fine arts.

[1] Dan. Starch, *Principles of Advertising*, p. 299, Chicago, 1923.

The evidence thus collected had to be compared with the results arrived at by institutions with aims similar to my own. It was necessary to pay the most careful attention to the work done by the Board of Trade Committee on Art and Industry in 1931–2 (Gorell Committee), the Board of Trade Council for Art and Industry since 1934 (Pick Council), the Design and Industries Association, and by the organizers of exhibitions such as the Joint Royal Academy and Royal Society of Arts Exhibition which was held at Burlington House in 1935. I also found it useful to keep an eye on periodicals dealing with modern design and recent books on the same and related subjects.

The present work is divided into two sections. Its first part is a bare account of the data which my informants gave to me. I was hardly ever able to check what I was told, just as the readers of this book will hardly ever be able to check what I am going to tell them. This being so I wish to stress here that the method I had to use for procuring information implies another limitation in my results. Manufacturers or shopkeepers may have made wrong statements, they may not have told me the exact truth or the whole truth. I need not say that in cases where I had the feeling that something of this sort was happening, I preferred not to use the evidence at all. Of the firms to which I wrote only about one-sixth refused point blank to see me. Their arguments are too characteristic of a still surviving attitude amongst provincial manufacturers to be left unmentioned. One firm producing electric fires simply stated: "This is the one department of our factory where we do not permit visitors." A firm in the toy and doll trade tried to justify a refusal by saying that the research in question would "enquire rather too deeply into our methods of conducting our business". A director of one of the best-known manufacturing and retailing firms in the English silver and electro-plate trade was more vague in his letter. He simply wrote that my "questionnaire was . . . in the nature of an inquisition with which (he) was not in sympathy". The reason why such industrialists were antipathetic to my research was given with disarming ingenuousness (and in remarkably awkward English) by the director of a well-known pottery firm with works in more than one town: "Our commercial arrangements are designed to advantage our factory in competition with a host of competitors, and I should not be prepared to have them brought into a general discussion." Though this statement shows an extraordinary ignorance as to the methods and aims of enquiries such as the present one, it is far surpassed by a passage from a letter that was written to me by an old-established firm manufacturing electric and gas fittings. It was worded like this: "We are . . . not interested in the question of public taste in design."

To these few cases I want to add the one instance of a manufacturer suspecting me of spying for a foreign power. On the whole, however, I found the industrialists whom I met much more responsive than might have been expected. A good many allowed me to examine their sales figures or showed me over their works. In spite of that, it may well be that my results in some trades are still one-sided and only partly true. For though I have found very marked differences in taste between London and Birmingham, and even more between London and the North, I was not able to do as much personal research in Lancashire, Yorkshire and Scotland as I had carried out in the Midlands and in London.

All these shortcomings of my report are evident. But I still hope that the facts which the following chapters contain may in the end be of some use, just because they are facts and nothing else. For, if I may point to a third criticism of much that has recently been said and written about the problems here in question, one should be very careful not to assert too much and to prove too little. So often we hear that the popular taste is not so bad as is generally supposed, or that designers in firms are not sufficiently well paid, or that artists as a rule are inaccessible to commercial propositions, or that manufacturers are not willing to risk experiments, etc., but hardly anybody has seriously attempted to examine what objective truth there may be in all such contentions.

This is the task of Part I of the present work. I should have liked to confine this part entirely to data and leave my personal judgments entirely to the second part. That, however, turned out to be impracticable. The problem of the prevailing taste in a trade, whether comparatively good or bad, is so closely connected with the conditions of design which I report that it cannot be discussed separately. In fact, part of my actual interviews with industrialists consisted of questions about why certain goods were as badly designed as they undoubtedly were, and why their artistic quality was not, or could not, be improved. The answers to these questions on matters of aesthetics often revealed facts about materials, processes of machine production and costs, which certainly belong to the descriptive part.

The whole of Part II is concerned with the conclusions to be drawn from the data in Part I. Why are the results in one trade better than those in another? What part does tradition play? What bearing has the payment of designers on the artistic value of the goods produced? What can be learnt from the experiments which firms have made in employing artists and architects? Are expensive things as a rule better designed than cheap articles? And if so, why? Preliminary answers to many questions of this kind form the basis for a series of suggested improvements with regard to the attitudes of the manufacturer, the architect and the artist, the organization of design departments in firms, the structure of art education, and practicable public and private propaganda.

But before entering into the discussion of all these problems, two questions of a philosophical nature have to be broached, though I know that it is only possible just to touch on them here. The one is: How can I, from my personal judgment, state what is good and bad in design? And the other is: What legitimate reason have we for feeling in the least concerned about better design? Why not leave it as bad as it is?

Any complete answer to the first of these questions would swell into a bulky volume on aesthetics. In our connection, it may be sufficient to refer to the succinct and careful delineation of some of the aesthetic laws underlying industrial design which Herbert Read has given in the introduction to his volume on *Art in Industry*.[1] Beyond that two very simple remarks are all we want. I do not believe myself to be infallible in my judgments on taste, but they may be taken as the conscientiously considered judgments of one who has tried to follow the ways of the Modern Movement in architecture, industrial art and the so-called Fine Arts on the Continent, and above all in Germany. The selection of objects reproduced on plates I–XIX may give a further indication of what seems to me to be satisfactory in taste.

Slightly less subjective is another argument which, in fact, may settle the matter, so far as we need it settled. Anyone who has ever been present at the meetings of the jury for a big art exhibition of really live importance will

[1] Faber and Faber, 1934.

have noticed that, if out of 5000 objects sent in only 1000 can be shown, there will be 500 which, from the first examination, are accepted by all experts, at least so far as they belong to the war and post-war generations. There will be 3000 at least, say 3500, which hardly one of the jurors finds himself able to take seriously. The remaining 1000 will be the border cases, the cases where arguments arise and convictions clash. In a discussion on 3000 objects of industrial art (e.g. in the British Industries Fair) the procedure would certainly be the same. There would be, from the first, unanimity as to two-thirds at least being out of the question, unanimity as to one-tenth being definitely good or acceptable—provided again the jury consisted of experts, that is men who are keenly interested in living art and industrial design, have seriously attempted for some time to evolve a power of discrimination, and are prepared to judge disinterestedly, objectively and conscientiously.

As to the border cases which between such a jury and myself may be left in abeyance, I am willing to give them the benefit of the doubt. For I do not think it is good policy to be as orthodox as some of the best English critics are at the moment. To abuse or ridicule any nostalgia for ornamentation can only deter people from studying the modern style and from trying to appreciate it. One tea-set may not be as perfect as another, but if it is decent in shape and decent in decoration, let us allow it a place among the sheep and not relegate it to the goats.

I need hardly add that I have tried not to mistake the modern for the beautiful, nor a fashion for that much profounder expression of a new period in the history of humanity which we call a style. Though the catchword "fitness for purpose"[1] has been misused by many supporters of the Modern Movement, and though, on the other hand, all the most recent English writers on art in industry enlarge upon the basic heterogeneousness of beauty and fitness for purpose, yet it seems as if to-day nothing of vital energy and beauty can be created unless it be fit for its purpose, in harmony with the material and the process of production, clean, straightforward and simple.

[1] A phrase, incidentally, that the architect Owen Jones used as early as 1852 (*The True and the False in the Decorative Arts*, published in 1863). Very similar is the wording in Pugin's *True Principles of Pointed Architecture*, 1841. Cf. also a lecture given by Frank Lloyd Wright in 1903.. ("The Art and Craft of Machine", reprinted in *Modern Architecture*, p. 15, Princeton, 1931.)

As to the second question underlying the whole of this book, we cannot so easily avoid discussion. Why should design be improved? Is good design worth a struggle? Personally I have no doubt that beauty, both of nature and of things made by man, beauty surrounding us in the streets, in the places where we work and where we live, beauty not only as a passing enjoyment of something outside our ordinary lives, but expressing itself in all the implements of everyday use, helps to make our lives fuller, happier and more intense. But that again can hardly be proved, though one negative test will be admitted by almost everybody, namely the degrading, debasing effect of dingy factories, dirty streets and dark dwellings on those who are forced to spend their lives amongst them. But in other respects the value of good design in industrial products is much more obvious. In a cardboard travelling-case made to imitate alligator skin, in a bakelite hair-brush made to imitate enamel—there is something dishonest. A pressed-glass bowl trying to look like crystal, a machine-made coal-scuttle trying to look hand-beaten, machine-made mouldings on furniture, a tricky device to make an electric fire look like a flickering coke fire, a metal bedstead masquerading as wood—all that is immoral. So are sham materials and sham technique.[1] And so is all showy, pompous, blatant design.

In fact all these horrors would scarcely have arisen, had not the industrial development of the nineteenth century deprived the poorer classes of so much joy in life. A splendour which reality does not concede is brought into our humble surroundings by meretricious industrial products, which achieve in permanence some of the elating effect that for a few hours is bestowed upon us by the Hollywood heroes' fantastic mode of life in the pictures. The pleasure which most people take in an entertainment so vicarious as the cinema, as well as the pleasure in vulgar and boastful design, is largely accounted for by the universal and irresistible longing for escape. Looked at from this point of view, the question of design is a social question, it is an integral part of *the* social question of our time. To fight against the shoddy design of those goods by which most of our fellow-men are surrounded becomes a moral duty.

So far we are on safe ground, and for the purpose of this book it is not necessary

[1] I admit that this principle cannot be pushed to its extreme. If we wanted to be really orthodox we ought to condemn all so-called crystal-glass (as Ruskin did) because it is not true crystal.

to go further. The question whether "Kitsch" (to use the untranslatable German word for any over-sentimental, over-sweet, luscious, in short "chocolate-boxy" production in art, music and literature) ought to be abolished wherever it looms in painting and sculpture is much more complex and intricate. It is somewhat doubtful how far ethical considerations can help to solve this. To a certain extent unquestionably—but is there really so much harm in the blatant blue sky of Naples or in the dazzling blonde perm of a wistful girl on cheap colour-prints in a workman's parlour? However that may be, the need of change here appears much less urgent than in the sphere with which my investigations are concerned.

One more word about the criticism which will be found in Parts I and II. Much of it is unsparing; but a book such as this would be completely useless if it tried politely to conceal the unpleasant truth. Things are extremely bad. When I say that 90 per cent. of British industrial art is devoid of any aesthetic merit, I am not exaggerating. A glance at the British Industries Fair or at any trade paper will corroborate this statement. However, it may be less insulting than it sounds at first. I do not know of any modern country where the majority of industrial products is not deplorably bad in design. So the aim of any campaign for better design can only be to reduce the percentage of objectionable goods from 90 to 80 or perhaps 75 per cent.[1] I firmly believe that this is worth fighting for. I hope that I have duly appreciated anything of good and honest design that I have seen, and I hope that my criticism will not be mistaken for superficial frivolity, but regarded as a serious attempt to help where help is needed.

[1] My estimates are in conformity with those of Mr John de La Valette, who writes: "I think that we can say that, roughly speaking, 10 per cent. of the manufacturers in every industry are fully in sympathy with the aims (of the Burlington House Exhibition). But using that 10 per cent. . . . you might hope, within a reasonably short time, to bring into line perhaps 20 per cent., or, if you are lucky, 25 per cent. . . ." To this I may add another statement which was given by one of Britain's leading textile manufacturers and which comes virtually to the same although it is more general in its wording: "There is no getting away from the fact that our standard of taste in this country is on a terribly low level, but then . . . it is not particularly high in any other country . . ." (F. J. Donald, Lecture on Furnishing Fabrics given to the Red Rose Guild and the Design and Industries Association, Manchester, 27 February 1935).

PART ONE
DATA

"*Sachenvoll, nicht wortgelehrt.*"
Herder, 1769

SURVEY OF TRADES INVESTIGATED

It must be realized from the outset that only a selected number of trades could be dealt with. Of the industries chosen about fifteen are described with a certain completeness, the remainder have been regarded merely as a subsidiary source of information. Six trades are concerned with the larger objects used in house furnishing—metalwork, electric and gas fittings and fires, furniture, furnishing textiles, carpets, and wallpapers. Six trades produce smaller objects for the house—pottery, glass, silver and electroplate, hollow-ware, products made from synthetic resins, radio cabinets. Two trades manufacture goods for personal wear—jewellery (including watchcases) and fancy leather. Dress materials are for simplicity's sake discussed together with furnishing textiles. A section apart is concerned with the design of motor-car bodies. Packaging and printing in general are not treated with any attempt at completeness.

METALWORK

The production of the firms which I saw comprises such heterogeneous objects as entrance gates, grilles, staircase balustrades, metal windows, commemorative tablets, door handles, name plates, metal bedsteads. Ten factories were visited. Owing to the variety of products I found more fundamental differences between them than in most other trades. Their methods varied between modern mass production and artistic "Guild" work; the number of employees ranged from 1500 to 100. One firm (1500 employees), mainly concerned with the production of fencing, hurdles and door furniture, had their own rolling mills, another (1100 employees) specialized in the manufacturing of standardized steel windows and metal roofing. The annual output is about 400,000 steel windows. In three medium-sized works (250–850 employees) up to two-thirds or three-quarters of the production consisted of locks, hinges, door checks, etc., i.e. wares giving little opportunity for considerations of beautiful appearance. The "Guilds" were smaller (100–200 employees) and based on the requirements of individual customers, one of them not limiting itself to metalwork but going in also for fibrous plaster, lead and stained glass work. One bigger factory (600 employees) had an even more composite organization, including departments for wood and stone carving, joinery and bronze-founding.

The methods of production also differed widely. In the fabrication of metal windows, railings or door handles no craftsmanship is needed. Once designed, the article is carried out entirely by machinery. On the other hand, in the manufacture of elaborate gates and grilles there is still a great deal of scope for the highly skilled craftsman, although in no case have I found a metal worker in a factory who did the designing himself.

This is done by staff designers, art directors and architects. As to gates, grilles, balustrades, etc., nowadays at least 50 per cent., but in most cases 80 or 90 per cent. of the inventive work rests with the architects responsible for the building for which the gate or grille is intended; and I was told more than once that this method was likely to become even more prevalent in the future. The "Guilds", proud of a tradition reaching back to the Arts and Crafts Movement, complained of this encroachment by the architect and ascribed it to the fact of his having less work and therefore more time for going into details. It appears doubtful whether this materialistic explanation is adequate; there are too many symptoms within the modern style which point to a legitimate widening of the architects' duties and rights. I shall have to return to this subject later.

Also, it rarely happens that the architect works out exact and detailed designs for a gate or a grille. As a rule he would hand over to the works' drawing office a rough sketch showing his general idea, and the rest would be left to the staff designer. Where the architect does not interfere at all, the active designing is done in the works. One mass-production firm, in which artistic metalwork is only a side-line, paid their designer £500 a year. Another comparatively big factory had two head designers with salaries of a little over £500 a year. In the "Guilds" much higher payments are made to those who do the creative designing. Usually one of the directors is responsible; he may even be one of the founders of the firm, at a time when the name of Guild was still more correctly applied than to-day.

In door furniture, the situation is quite different. Here staff designers, paid at a rate of £250–£350 a year, are originating new patterns. But their freedom of invention is considerably restricted by suggestions from the sales departments and travellers, as to what competitors have successfully produced and merchants repeatedly demanded. The salaries are small, but it has to be borne in mind that the need for new designs in door furniture is nowhere large enough to justify the expense of a more highly paid employee. Moreover the production of handles, levers, etc. was, in all the firms that I saw, only a side-line, occupying a comparatively small proportion of staff and plant. This is of importance; for in the steel-window works referred to, i.e. a factory where most of the production is connected with the activity of the designer, a salary of £800 is paid to the man in charge of the drawing office.

As to the *artistic character* of the work produced by the firms which I visited, first a division has to be made between period design and modern design, everything being regarded as period that follows the lines of the historical styles, everything as modern that is independent of these. In gates, etc. I was told that the Modern Movement began to penetrate about six years ago, in door handles, etc. only four or five years ago. But in gates and grilles even those firms which are run on modern mass-production lines still produce 50 per cent. in more or less traditional forms. In one of the "Guilds" the corresponding figure was as high as 75 per cent., and in Mr Dowling's book on *British Industrial Arts* (Lewis, 1934) there are fourteen pages illustrating gates, grilles, etc., of which only one is what I should call modern in design.[1] The situation in door furniture is just the opposite. There is hardly any demand for period knobs, finger plates, etc., almost every pattern on the market being modern, or rather "modernistic", i.e. bristling with unexpected jags and angles.

How can this be explained? The usual argument—that the public wants it—does not hold good here. For the public is hardly ever asked. Almost all the trade is done through builders' merchants. It is they who dictate the taste. Such at least was the conviction of the manufacturers whom I met. There was no unity of opinion as to whether the distributor represents the true taste of the

[1] In only one firm was I told last year that the bulk of their artistic metalwork was modern. However, the same firm laid great emphasis on the fact that the demand for "period" forms has been increasing again during the last twelve months or so. Also the photographs shown to me as illustrating recent work of the firm were as conventional in character as most of the production of the other factories.

consumer. Some asserted that the merchants as a rule have "their fingers on the pulse of the public", some distinctly denied this, one in particular who strongly disliked what he called "futuristic" forms and explained—an argument worth considering—that Victorian handles were at least comfortable to grasp and could not be drawn without a certain knowledge of design. But he pronounced himself powerless against the merchant, who would stop buying from him unless he offered modernistic patterns. Here for the first time we are faced with the important question of the distributor's influence on the design of products.

Another point not without bearing on design is price. In one of the "Guilds" it was suggested—wrongly, I am sure—that the demand for simpler and straighter forms of modern gates and grilles was due to the customers' lack of money. This implies, as in fact I was once told, that more elaborate decoration would be preferable and might return to favour as soon as the economic crisis in Europe and elsewhere should be passed. Although this materialistic conception appears to be corroborated by the return of reproduction forms since the economic recovery of the last two years, its fundamental falseness becomes obvious as soon as one recalls the fact that the Modern Movement started on the Continent before the war in a period of far greater prosperity than that in which we are living now. So if there is any relation between the modern style and that shrinkage of private wealth which may well characterize our century, it can only be due to a deep-rooted "Zeitgeist" expressing itself in art as well as in economics. On the other hand, it is undeniable that if the builders of housing estates with several hundreds of houses do not wish to pay much for the necessary door furniture, then elaborately decorated design, which would increase the costs of tools, is out of the question. Now, certainly modernistic patterns require less complicated tools than those of Victorian taste, masquerading as Baroque or Adam style. But, if economy were the only governing factor, would it not be cheaper still to do away with the cubistic jags also?

As to the *artistic quality* of English metalwork to-day, my result may appear strange at first sight. Many of the steel windows used in new houses are almost perfect in proportion and clearness of expression. And this is true not only of special designs, but also of the best-selling types in the firm which I saw. These pleasing proportions, well suited to the most varied types of houses, were attained by a system of progressive standardization, eliminating more and more of the once existing forms. The standards were worked out by a council of architects in the presence of the designer of the firm in question. He himself had been trained as a constructional draughtsman in an architect's office. A good many architects, I was told, objected for almost as long as ten years to the use of such standards, as they did not want to be restricted by any agreement between manufacturing firms in designing individually proportioned window-panes. In the meantime many have realized that this standardization has not only kept prices down, but has also resulted in a considerable improvement of the average standard of window design in cheap builders' houses. The fact that here co-operation between enterprising and enlightened mass-production firms and a few technically interested architects has led to such excellent results should be borne in mind.

While, as I said before, in door furniture the atrocities of modernistic "jazz" patterns rage, in entrance gates to parks or public buildings and offices hardly anything truly modern (or modernistic) has so far been developed. This applies to the Continent as well as to England. The majority of gates, grates and grilles are still designed on bygone standards, as is proved by some of the figures quoted above. What is the explanation of this state of affairs? One reason at least may be hinted at here: the modern movement has so far found a much more genuine and complete expression in unadorned shapes and surfaces than in any kind of decoration. We are willing to believe in a new beauty of our century, when we recall the pure and clean-cut shape of a liner, an aeroplane, a racing car. Any decoration applied to them would be ludicrous. Hence most modern architects have almost entirely renounced ornament, and consequently keep aloof from purely decorative work. When they are asked to design a monumental entrance gate, most of them feel at a loss. And, on the other hand, those commissioning the bronze doors of a town hall or a bank usually insist upon getting some elaborate design, not a piece of work whose effect lies in the pureness of proportion and the beauty of the material alone. In these circumstances, what usually happens in England is that the architect either takes refuge in some slightly modified neo-Classic or Georgian, or else leaves it to one of the "Guilds" to design something that will satisfy the town council or the bank authorities. The "Guilds", however, faithful to their origins in the Arts and Crafts Movement, find it difficult to change their outlook, though their methods of production are of necessity far from comparable to any metal workshop of the mediaeval, guild-organized centuries. While the change from the art worker's shop to the factory had to be made, the corresponding step from a handicraft style to a machine style in artistic metalwork remains to be accomplished.

True as this is of gates, etc., it does not hold good for another line of production in the factories of the "Guild" type. I am thinking of commemorative tablets, name plates, etc. Here the Morris movement seems to have improved the common taste so profoundly that now the standard of English lettering is above that of any other country. Owing to the work of men like Emery Walker, Edward Johnston, A. H. Mason, Eric Gill, anybody now attending English art schools will get a sound introduction into the principles of lettering. The influence exerted by such outstanding artists has revolutionized the trade in its entirety. Staff designers and even young draughtsmen, trained part-time in art schools, can here show what they are able to do when they are left alone to make something not "period" or "modern", but clear, well balanced and legible.

ELECTRIC AND GAS FITTINGS

In this section, heating apparatus as well as lighting fittings will be discussed. It appears inadvisable to separate the two, because as a rule the same firms produce both and the same methods are used for evolving new designs. Yet there is one essential difference between a ceiling fitting and a fire. Fitness for purpose is the one decisive criterion in judging the value of designs for electric or gas fires; a pendant or a table standard should have certain decorative qualities with a purely aesthetic appeal as well. The buying public realizes this and insists, in purchasing a fire, on a particular maker's name, while it looks on a ceiling fitting as an art object. Nobody would say that he prefers a certain firm's pendant fittings to those of any other, but it is quite usual to hear similar statements on Portcullis, or Radiation, or Magicoal fires.

Consequently, the history of electric and gas fires[1] yields results quite different from those to be derived from a history of electric lighting fittings. Gas lighting

[1] On gas fires, cf. L. Weaver, *Gas fires and their settings*, The Fanfare Press, 1929, and T. W. Coghlin, *Architectural Review*, vol. lxxv, 1934.

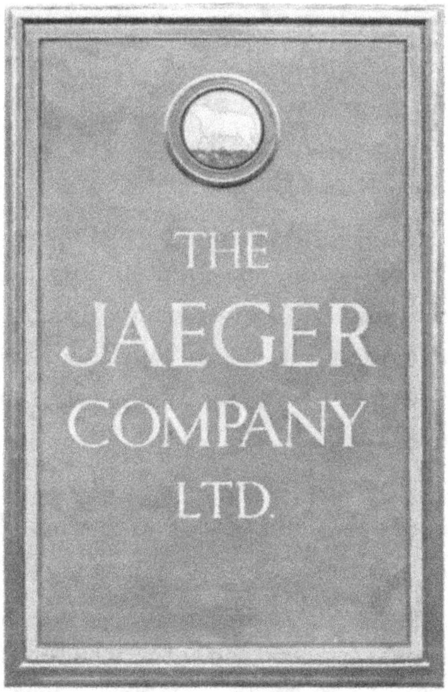

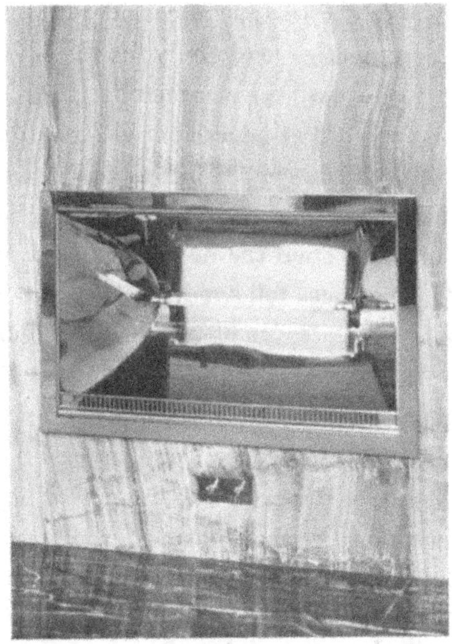

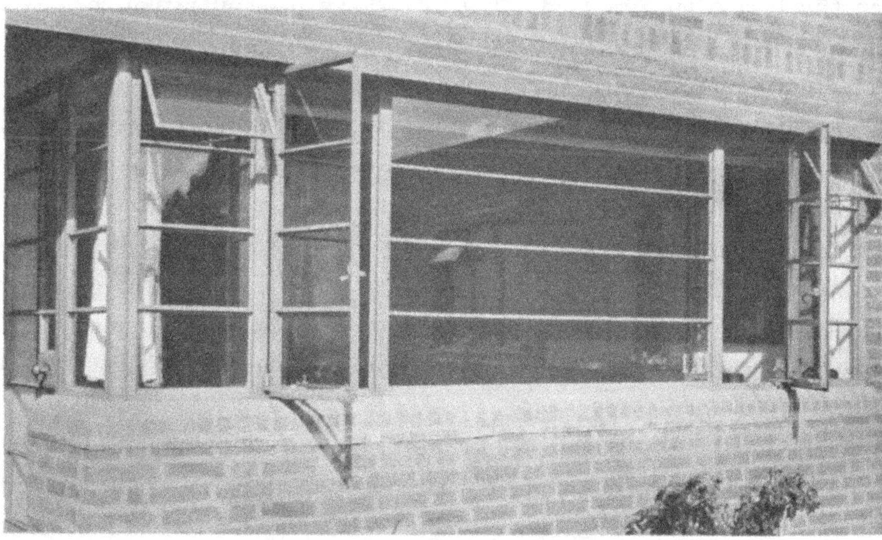

I. *Name plate designed and made by the Birmingham Guild.*
Electric fire of exemplary design. (Ferranti Ltd.)
Standardized steel windows. (Henry Hope & Sons Ltd.)

fittings can be left out of consideration. They are nowadays—except for the very simplest models—made only for display in gas company showrooms. The earliest electric and gas fires were not designed at all, at least so far as artistic design goes, i.e. no special consideration was given to the problem of appearance. A gas fire was black and grim, as inconspicuous as a gutter, and an electric fire was a round copper bowl, not bad-looking, but obviously conceived as a heating machine exclusively. Then the possibility of applying art was discovered, and a phase of artistic outrages followed. In gas fires, some timid period ornaments were spread over the surfaces; in electric fires, Tudor and Adam surrounds flourished, and all the characteristics of the traditional solid-fuel fireplace were introduced into the designs. Even where something new was aimed at and the style of the Arts and Crafts was adopted, the result seems hybrid now. An interesting illustration of this stage in the evolution of modern heating apparatus is the introduction of the flame-shaped radiant into the design of gas fires. The clay cylinders which keep the single gas flames together and separate them from each other were (according to what I have been told) adorned with curved fretwork not for efficiency's sake, but because of the pleasing, appropriate, and so to speak, symbolic appearance which this decoration provided. It soon became so popular that for years the public took it as a technically necessary part of the gas fire.

The more recent and decisive change of outlook is connected in electric fires with Ferranti's, in gas fires with Bratt Colbran's. The first fires based on the idea of the rectangular incurved reflector were brought out by Ferranti's in 1926. The appearance of these was gradually improved by casting off some unnecessary frills, until in 1931 a model was achieved which is still to-day one of the most satisfactory on the market. At about the same time, or more precisely in the autumn of 1932, the first Portcullis fires were shown; gas fires in which the radiants form a grate or gridiron of Portcullis shape, an artistic improvement of great consequence.

Beyond this historic parallelism, there are two more important points in common between the new departures of Ferranti's and Bratt Colbran's; one of them concerns methods of design, the other sales. The reasonable and attractive radiant of the Portcullis fire resulted, according to the information which I received from the works, from experiments by engineers to attain an increase in the heat output. However, since they considered this problem seriously and comprehensively, and endeavoured to carry it to a solution as near perfection as was within their power, they arrived at the same time at something that was aesthetically satisfactory. The engineers' final proposal was shown to the artistic designer of the firm, and he was honest and appreciative enough to say at once that he could not have created anything better himself. The origin of Ferranti's incurved rectangular radiator is very similar. The late Mr De Ferranti was an engineer of wide experience and considerable imagination. The conception and development of the new idea was largely his work. At that time the firm had no special artistic designer. This applies not only to the 1926 but also to the 1931 model.

To find out how far the conditions which I had met within these two firms are typical of the whole of the trade, I interviewed four more manufacturers of electric fires, four of lighting fittings and three of gas fires. The total evidence thus collected embraces, in electric fires, data from two of the biggest producers in the country, one of whom has about 7000–8000 employees, and the other about 4000. In the first case about 2000 are working in the department where the electric fires are made, the others being employed in the production of telephones, radio and other electric apparatus. In the other firm only a relatively small proportion of the employees are concerned with fires. The three other works employ 700 (fires and fittings), 60–70 (fires and fittings) and about 200 (fires, fittings and domestic appliances). As for electric lighting fittings, three of the firms which I saw have been mentioned already as makers of electric fires, the fourth is one of the pioneers of the modern movement, a comparatively small London house with a staff of seventy. In gas fires, my results are founded on statements made by one of the biggest manufacturers in Britain (with 2500 employees), another old and well-known factory (with about 900 employees), and a third which, though equally big, employs only about 100–200 workers in its gas fire department.

As to the development of new designs for fires, the following facts were given to me. It is not usual to employ a designer of recognized artistic abilities and of artistic training, but either a simple head draughtsman who is paid about £250–£400 a year for putting on paper what is suggested to him by directors and sales managers, or alternatively a man who has gone through a technical school or technical college and happens to possess some artistic talent or ambition as well. The creative work is then done by the directors or one of the directors. If he is really interested in art (as, for example, is the case in one of the firms which I know, where the director has been through a progressive art school on the Continent, keeps abreast of all new developments at home and abroad, and willingly accepts fresh ideas from good and adventurous architects and designers) the result will naturally be very satisfactory and better than if the director is only technically competent. But even in the latter case, as a rule, nothing hideous is turned out, though probably much that is insignificant.

On the other hand, it was pathetic to see the pride of one manufacturer of fires and fittings who showed me what he had done for the sake of art. He was an engineer and did the basic designing himself. He had, however, made it a rule to call in an artist from London and to leave it to him to embellish shapes which otherwise might appear too technical. The result was that a good deal of what the engineer had done, quite soundly and neatly, was obscured. No wonder; for the additional application of art to something not visualized in terms of art right from the beginning, or the application of one man's art to an object created by another man in terms of his conscious or unconscious art, can never be aesthetically successful.

Broadly speaking, artists with a thorough training and a genuine interest in technical questions, such as I have found in two or three firms (in the case of two, they were members of the board of directors), are certainly able to improve on the artistic value of fittings and fires, but the most satisfying results in this trade seem to have been achieved by engineers and manufacturers working all by themselves and unaided by specialized artists. I do not know of any electric or gas fire designed wholly by a free-lance artist or architect; so this case has to be left out of consideration.

Some examples from another industry akin to the electric and gas fire business seem, however, to indicate that this might also be a promising method. In Herbert Read's book on *Art and Industry*, it is said that the design of the Swedish Aga Cooker was due to Dr Dalén, an engineer of international standing. I asked Messrs Svenska A. B. Gasaccumulatoren to give me some more information, and in their answer it was stated that Dr Dalén has in fact "taken an active part in putting different details into proper shape. His assistants in the drawing office have been responsible for the design of all the details. For a few details of seemingly minor importance which have been difficult to give the shape that appealed to us (the firm), an architect has been employed as adviser." Equally interesting are the statements which I received from the manufacturers of the Esse cookers and stoves. While the design of the cookers is evolved by team-work, based on the ideas of the managing director, incorporating suggestions from (technical) designers and salesmen and carried out by the works' drawing office, the new B. J. Esse Heating Stove, of a remarkably satisfactory appearance, was designed by Mrs Betty Joel, a distinguished designer of cabinet-work.[1]

To return to the story of Ferranti and Portcullis fires, I have said that another similarity between the two is to be found in the development of sales. Both the Portcullis radiants and the Ferranti reflectors were at first too new and unusual to be commercially successful. It took about a year before the retail buyers, the municipal gas companies (and consequently the public) could be converted to the new radiants, and it took well over four years before the same happened with the Ferranti reflectors. Then at last the market was conquered, and—this is the best, and in fact the only, proof of such a success—copies and imitations were brought out by rival firms. Now the biggest English manufacturers of gas fires have given up the flame-shaped radiant, and almost all firms which produce electric fires turn out something on the lines of the Ferranti reflectors. In this context the following facts may be of some interest and are certainly encouraging for other manufacturers who feel inclined to embark on some new adventure. In a large department store in the Midlands, I was told that the Ferranti fire of the reflector type, in a size which sells at 25*s*., is the store's best seller, and Messrs Ferranti told me themselves that their sales of this one model in this one size come to 18 per cent. of their total sales of electric fires.

[1] In this connection an article in *Trend*, vol. i, p. 36f., 1936 is of interest in which the evolution of a refrigerator design for Sears Roebuck's, the American mail order store, is described. The artist who is responsible for this unusually clean and yet distinctly smart design is R. Loewy, one of the most famous American "stylists". In the same issue of *Trend* there are also two pages on a new and remarkably original electric iron of H.M.V. This, I was told at the works, was developed in all its main points of appearance by the engineers and the draughtsmen on the staff.

However, to limit the general importance of this gratifying fact, it must be added that it applies, in gas as well as in electric fires, only to medium-price products. The most popular cheap models in both trades are made in vitreous enamel, and designed in simple forms, sometimes more, sometimes less, attractive. Yet being—as all fires should be—first and foremost conceived as useful machines, their appearance is as a rule not offensive and often in fact quite satisfactory. No effects of sham grandeur and meretricious glamour are attempted.

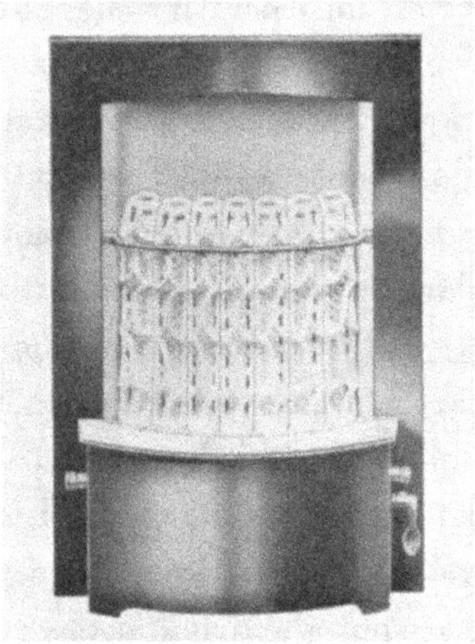

II (*a*). *Best selling type of inexpensive gas fire. (Parkinson Stove Company.)*

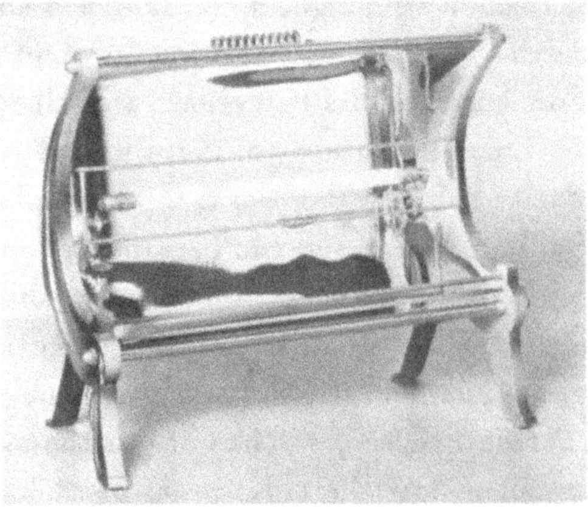

II (*b*). *Best selling type of medium-price electric fire: Ferranti's reflector model. Selling price about 25s.*

This, however, cannot be said of the most expensive type of electric fire. It is a fact that in all the department stores where I enquired, in the high-price range the "flickering" imitation coal fires of some brand were far and away the best sellers. In a big shop which I visited in the Midlands they accounted for 90 per cent. of the total sales in their price range; in the works manufacturing the most popular brand, this type represents 75 per cent. of the output. Ferranti's, who for a long time had not touched this type, have now also started to make it, and the output of this one type amounts to about 15 per cent. of their total production.

This is somewhat harrowing from our point of view; and it is scarcely sufficient compensation to find that the cleanest and most logical solution of the problem of heating a room by electricity or gas—the panel fire—is also becoming increasingly popular. Of this development stores cannot give figures, because most of the panel fires are not sold to private customers but direct by the manufacturers to builders' merchants, from whom they go to architects and builders. The panel type is supposed by now to account for 20–30 per cent. of the total demand for fires.

In electric lighting fittings the situation is not dissimilar. Again a historic survey of the evolution which has occurred during the last ten or fifteen years

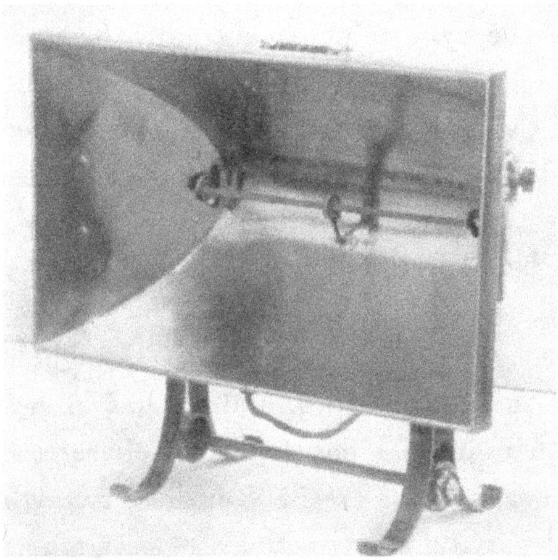

II (c). *Ferranti's first reflector fire, 1926. The beginning of sound design in electric fires.*

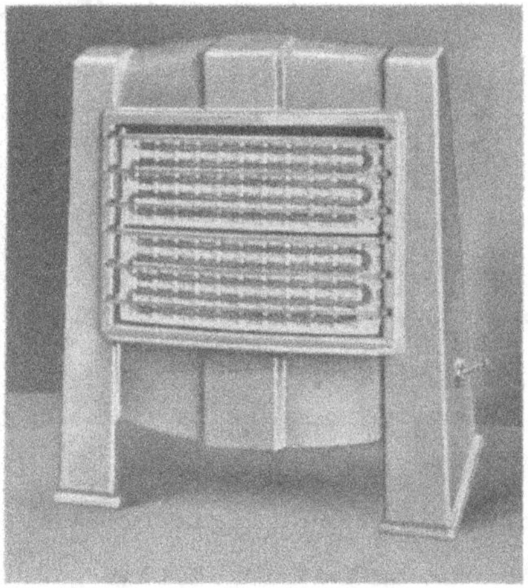

II (d). *Best selling type of inexpensive electric fire (G.E.C. vitreous enamel finish).*

will be instructive. In England there were hardly any but period fittings between the war and, say, 1925. It was the ominous Paris Exhibition of 1925 which must be held responsible for the introduction of bad "Modernism" into the trade. Jazz fittings, i.e. fittings with pseudo-cubistic, angular glass panels, became the fashion. Similar spiky forms had had a great vogue in Germany about 1922–3 in connection with the style which dominated in "Expressionist" painting during these most restless years of post-war change. In Germany, however, a reaction against the vulgar commercial caricatures of this type set in almost immediately. The Bauhaus brought out some simple spherical fittings in 1923. Since then, unadorned lighting fittings, of this as well as of tubular and square box type, steadily gained ground. For a short time they seemed to have almost completely defeated the more elaborate shapes. The final result, however, was a compromise, and chandeliers with three, four or six arms carrying tubular or bowl-shaped holders for the bulbs became the favourite kind of ceiling fitting. In England, box fittings were introduced about 1927, tubular and spherical fittings slightly later. The merit of having achieved this great improvement is due to a few adventurous manufacturers such as Troughton & Young, and Best & Lloyd. But they were still preachers in the wilderness when I began my research. The most popular type of ceiling fitting then was the modernistic panel-pendant with jagged orange glass panels. This was the type which went into the tens of thousands of semi-detached houses, where sometimes (as I was told) builders do not want to spend more than £3 on all the electric lighting fittings of the house. It was also the type which to the head of the electric department in a London store appeared so much more "homely, comfortable and harmonious" than box or bowl fittings. These, incidentally, he appreciated only for bathrooms and landings.

Fittings of this kind were (and are) usually the work of the same draughtsmen who are responsible for the design of inexpensive fires, and, as has been said above, such draughtsmen are as a rule in an inferior position. It is characteristic enough that amongst the firms which I saw it was precisely in those two where the best lighting fittings are produced that different methods of designing had been adopted. In the one which I mentioned before in connection with fires the manufacturer himself does the creative work, sometimes by himself, sometimes aided by his head designer and sometimes by modern-minded outside architects or designers. He assured me that he would be happier if he could produce nothing but really good fittings. But this, he said, he could not afford to do. Every now and then cases arise where he has to garnish a new model, because it would appeal to too small a market were it to remain as straightforward and functional as originally conceived. In these cases the catalogue of the 1925 Paris Exhibition is a welcome and inexhaustible source of sham splendour. In the other firm which is of importance in this connection, the owner, a young and enterprising man, some years ago engaged a designer who had come from the Royal College of Art, and made him his Director of Design. All artistic questions were left to him, with the result that the firm within a few years gained a leading position in

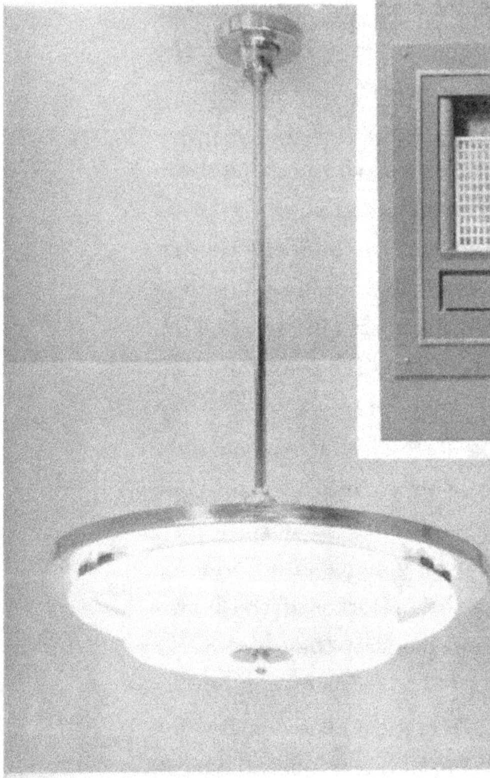

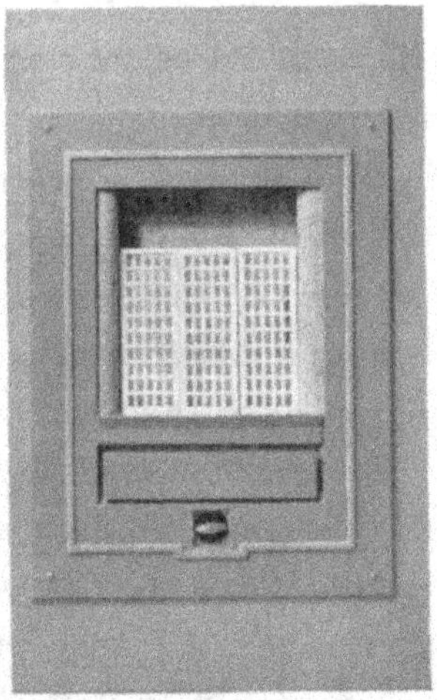

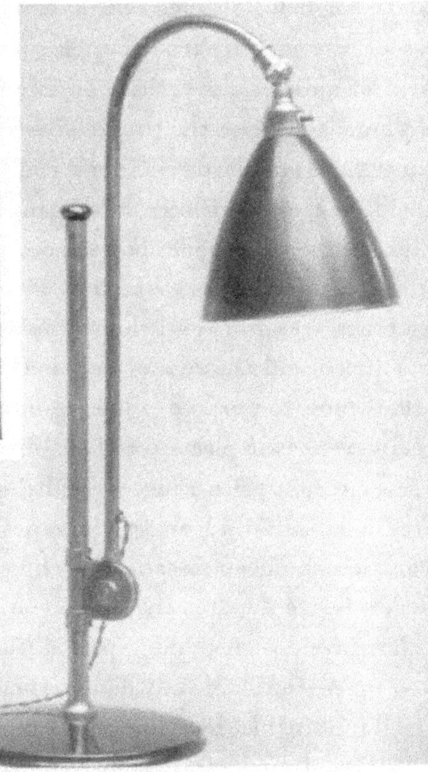

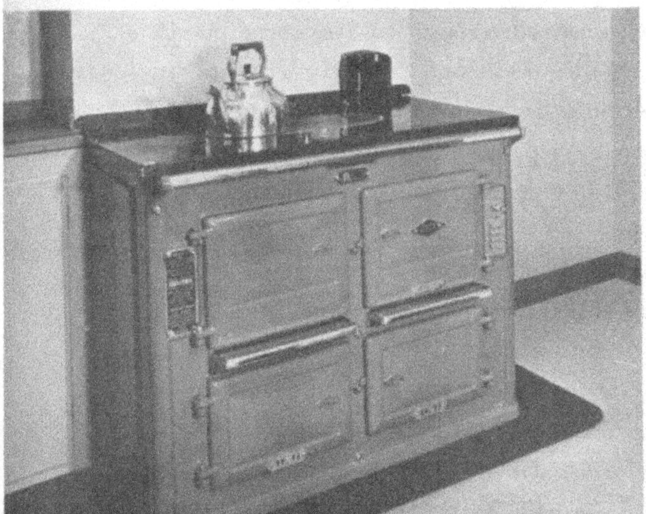

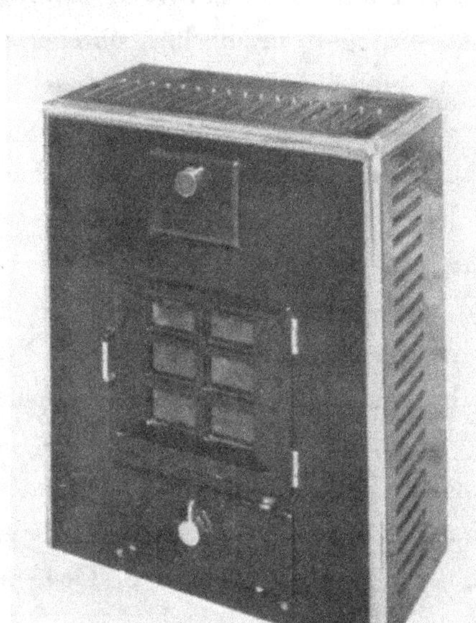

III. *Ceiling fitting by Best & Lloyd.* *A Portcullis gas fire.* *The "Bestlite" made by Best & Lloyd.*
An Esse cooker. (Smith & Wellstood Ltd.) *The B. J. Esse stove, designed by Mrs Betty Joel.*

modern English lighting. This firm, incidentally, does not manufacture either the glass or the metal parts of their fittings, though they are responsible for the design of most of the articles on sale in their shop.

Whereas in 1933 the progressive firms still had almost the whole of the trade against their standards and aims, considerable improvements have since then been made. The General Electric Company kindly informed me that in 1934–5 it was no longer the jazz model but something similar to the German chandeliers with upright tubular holders which had become their best seller, and in 1935–6 public preference was for bowl-shaped models of German or Swedish character. This development once more may be taken as a proof that the efforts for the promotion of the fitness-to-purpose principle have after all not been wasted.

However, as in electric and gas fires, this gratifying state of affairs exists only in the medium-price range of lighting fittings. Once you go into the multiple stores, you still find orange jazz designs or a nasty so-called "modern-floral" decoration on cheap so-called parchment shades. And once you get into London's most exclusive department store, you will notice that the majority of the lamps on show are designed along period lines. The buyer told me that they had never given up period lamps, although about two years ago the victory of modernism seemed irresistible, but that now (during the last twelve months) the demand for reproduction work was evidently going up once again.* I shall refer to this alarming statement again in another context. Corresponding to this, I found that the best selling models in two other department stores, one in Oxford Street and one in the south-west of London, were relatively cheap gilt and painted wrought-iron chandeliers in period forms. Not that period fittings play a really important part in the trade in general. Two firms gave me 25 per cent. as their ratio and said that, on the whole, London was less partial to reproduction work than the North and Scotland.

In expensive ceiling fittings, the situation is more complex. As was the case with grilles and gates, a large proportion of the most expensive work here, such as the fittings for cinemas, hotels, public buildings, etc., is done to architects' designs. This entails of course a great variety of different types. But period forms are rare. Somehow it seems as if for electric lighting not even a very conventional architect thinks in Tudor or Georgian forms. However, generally speaking, it

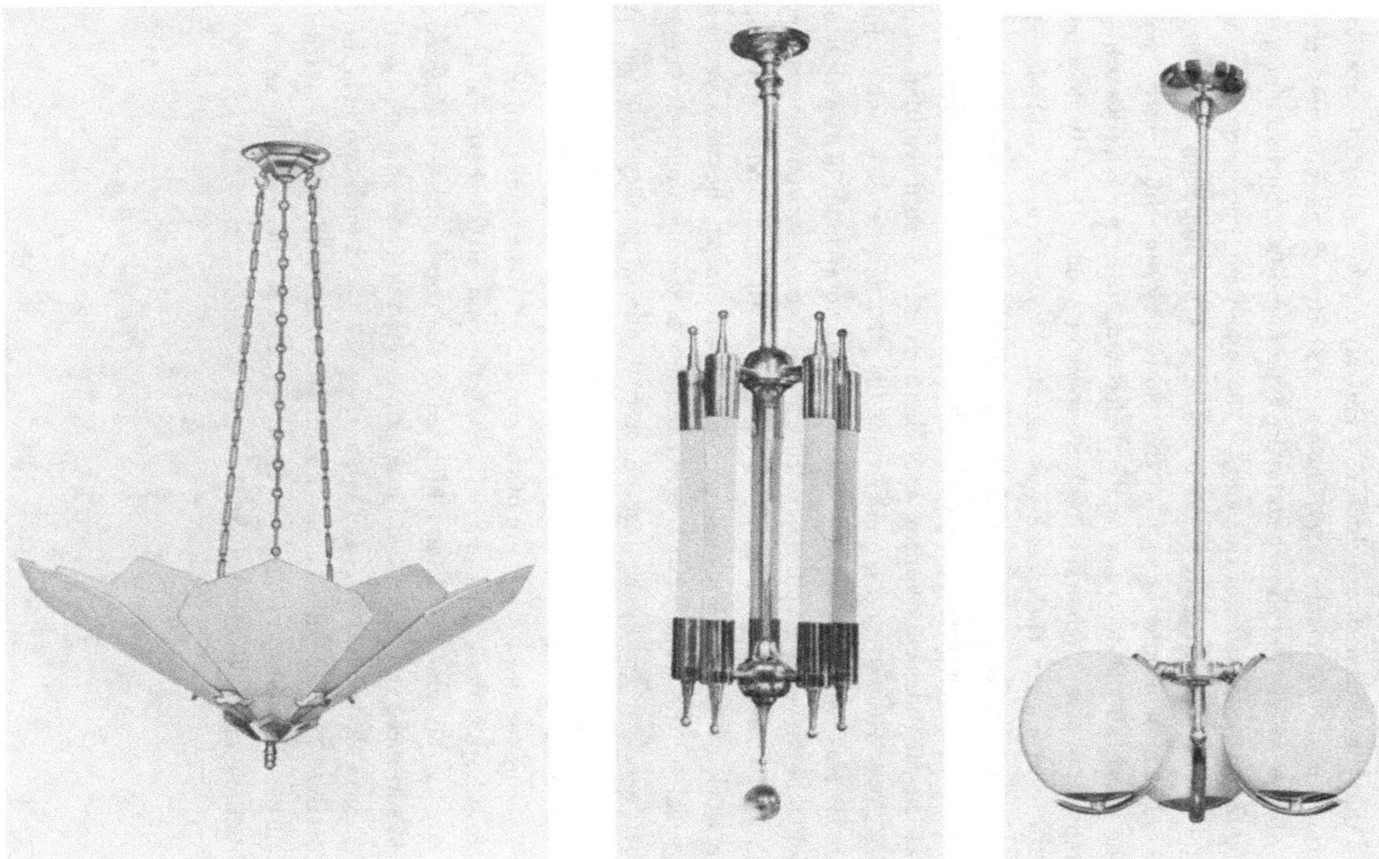

IV. Three best selling electric ceiling fittings. The first was one of the most popular types in 1933–4, the second in 1934–5, the third in 1935–6. (By courtesy of the General Electric Co. Ltd.)

must be said that dignified designs which rely entirely on the beauty of materials and proportion are also extremely rare. The vast majority of the more elaborate fittings are of that bright and showy stalactite type found in suburban cinemas.

In looking for an explanation of this and, at the same time, for the reason why the most expensive electric fires are so discouragingly wrong in design, one has to face the fact that most people, as soon as they go to buy something costly, will throw to the winds all the doctrines of simplicity and fitness for purpose which they are prepared to accept when buying less expensive goods. Directly they feel that they can insist on Art, their taste goes astray. Now, in the case of electric fires, which—as we have seen—are considered by the public primarily as technical apparatus, Art means for these people the clever tricks of the engineer that make electric heating elements appear as flickering coal. Should we blame the engineer for this? His attitude is understandable. He is just as proud of his ingenious gadgets as was the patient sixteenth-century craftsman of the skill of his hand when he carved a crucifixion with fifty-two attendants out of a cherry stone. Both achievements have nothing to do with aesthetics. And it may well be that the work of the modern engineer appeals for almost the same reasons which made these little wonders of the Renaissance so widely admired; it may well be that here we are up against an eternal human quality and not an atavism.

It is different with decorative ceiling fittings—and yet whom shall we blame? The public craving for the atmosphere of the cinema? Or the manufacturer? Or the architect who has designed them? I am afraid that his aesthetic failure is caused by exactly the same problems which led to the perseverance of reproduction forms in entrance gates, etc. Again it must be stated as a fact that most of the modern architects obviously feel at sea when they are asked to design something purely decorative. There cannot be any doubt that this inability is the outcome of some fundamental quality of our age. However, any farther-reaching speculations on this topic must be left to Part II.

FURNITURE

The fact that electric and gas fittings are products so closely related to engineering determines the conditions of design in these trades. The situation is quite different in the cabinet-making industry. As the problem here is rather involved and varies surprisingly according to the character of different firms, I collected information from as many as fourteen manufacturers of wooden furniture, three manufacturers of steel furniture and seven stores.

As to wooden furniture, two of the factories which I saw were of a general mass-production type, one of them employing about 3000, the other about 2000 workers. The average retail price of their suites is about £12–£15. Two were specializing in certain kinds of mass-produced furniture. One of them (500 employees) deals mainly with a patented type of chair, the other (200–250 employees) with inexpensive deal and beech furniture. Three works—one in the Midlands (about 300 employees), one at High Wycombe (600 employees) and one in Scotland (350 employees)—turned out medium-price goods (£15–£30 a suite). One firm was chiefly known for its show cases and fittings for shop windows; in another wooden bedsteads were produced as a side-line, the bulk of their trade consisting of light metal products. In addition I visited five firms of a more exclusive character. One of them was particularly interested in the execution of individual orders for the complete furnishing of houses or flats and produced only a few "stock lines", the second was among the makers of the most expensive suites to be seen in the big stores, the third (200 employees) was turning out at the time, besides more conventional high-quality suites, one range of expensive, but machine-made modern furniture the designs of which may be regarded as the most progressive in the country. The remaining two manufacturers have set equally great store by good modern design. One of them is in London and also runs a well-known store for the sale of fabrics and objects of industrial art and handicraft, the other (350–400 employees) has a factory and a shop in London and workshops in one of England's most charming villages. My evidence from shops came, apart from the London shops just mentioned, from one big furnishing house in London, from the furniture departments of three of London's largest department stores, and from two department stores in the North of England.

The bulk of the furniture trade nowadays is, I understand, in the hands of big factories well equipped with modern machinery. Nevertheless, much scope is still left for the owner of a small workshop, working alone or with a few apprentices. In the provinces some of these small producers are still craftsmen in the genuine sense of that word. In London the development of big business has deprived most of the small Jewish workshops in Shoreditch of their creative initiative and reduced them to the routine work of supplying parts, such as chair backs, drawers, legs, to the large firms who do the assembling. The small unit has not been driven out, but forced into sheer mechanical repetition work.[1]

Another peculiarity of the trade is that a considerable section of it is localized in one small town, High Wycombe. There are about 150 firms at Wycombe giving work to about 8000 employees. Only four or five of these firms, however, are considered outstanding in the trade. No more than two manufacturers employ more than 500 men, hardly more than four 250 and more. The average number of workers in a Wycombe firm may be about forty. The time is past when Wycombe made most of the chairs used in Britain. To-day all the bigger firms make "suites" of some sort.

The quantity of machinery used varies considerably in the different firms. While the coarse preliminary processes are everywhere—except in the workshops of a few purist craftsmen still carrying on the Gimson tradition—done by machines such as circular saws, the dovetailing, moulding and carving is done by hand in the most exclusive workshops which I saw. But even the firm which, as I said, specializes in the furnishing of expensive modern houses and flats admitted that all the moulding in their shops was done by machinery.

Expensive suites or single pieces are as a rule produced in quantities of two, three, six or up to twelve at a time. Batches of 100 and up to 250 pieces are usual in medium-price goods. The factory of patent sprung chairs which I visited turns out about 1000 a week. Of their best seller, they produce 3000 or 4000 in a year. Similar figures were given to me by the manufacturer specializing in low-price deal and beech furniture. The biggest mass-producer in the country, however, says that he sometimes makes more than 500 a week of one suite. The method used in this firm is to put through large quantities of parts, e.g. backs

[1] Cf. *The New Survey of London Life and Labour*, vol. ii, 1931.

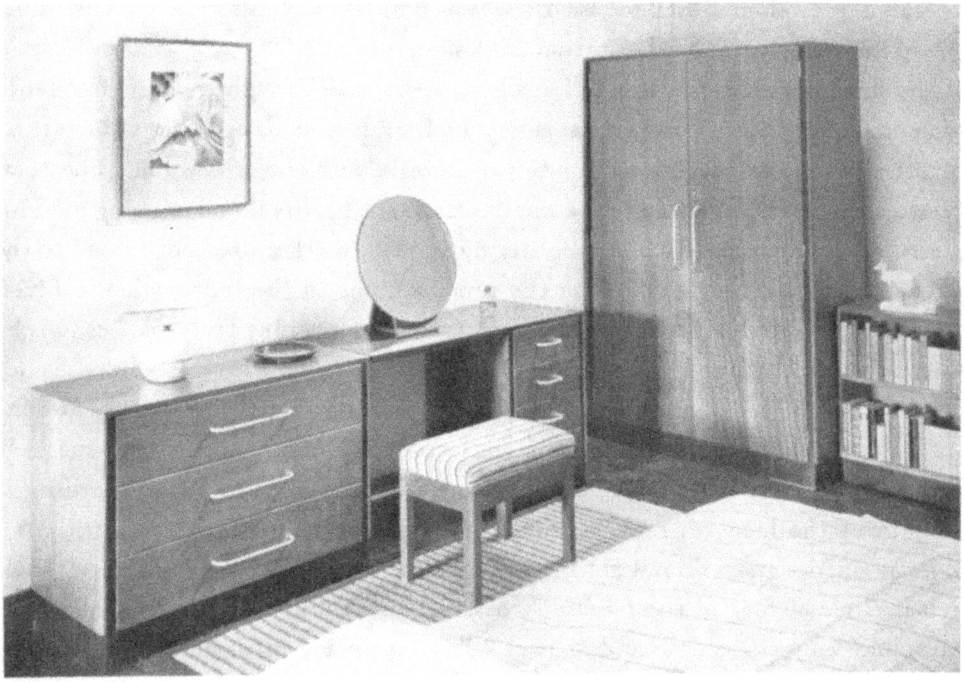

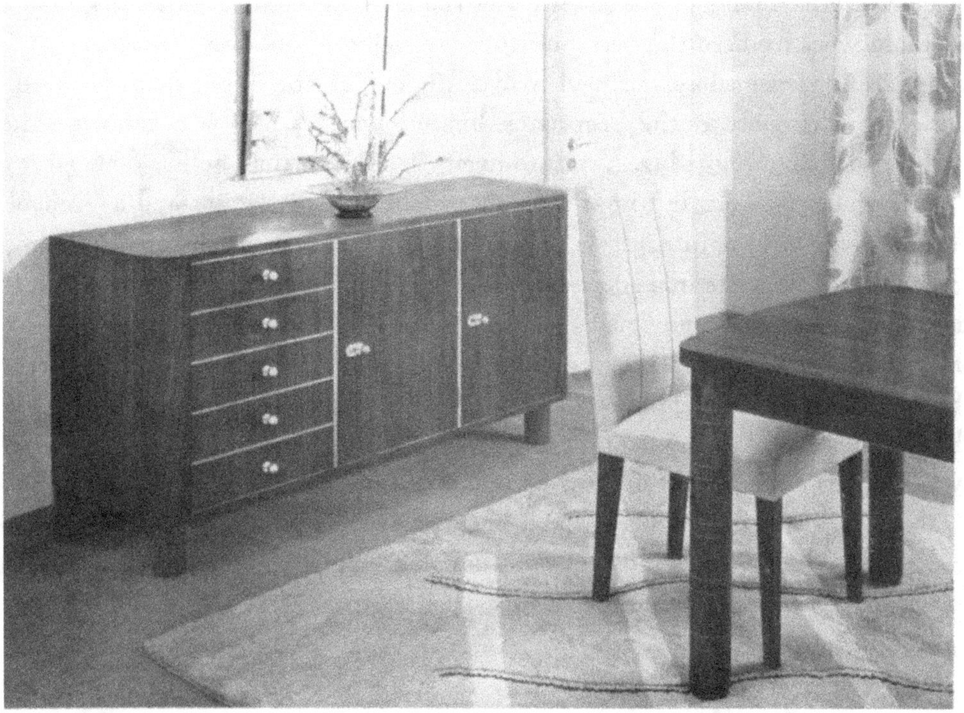

V. *The "Finstock" bedroom in natural mahogany and the "Woburn" dining room in rosewood, designed and made by Gordon Russell Ltd. On the floors hand-tufted rugs designed by Marian Pepler.*

of drawers, or doors, and to use these for building up different models. This system naturally entails a limitation in design.

The designing of furniture is as a rule done by staff draughtsmen or free-lance commercial designers. The rare cases in which architects design the whole of the furniture for a wealthy man's house can here be left out of account. The commercial designer sells mainly to the small manufacturer, his sketches being paid for at a rate of one guinea each or slightly more. His market does not extend to the big firm with a drawing office on the premises, nor to the independent cabinet-maker who still relies on his own talents trained in evening trade classes, or else on copying. Copying, I was told, is also one of the main sources of supply for the design of Wycombe products. Not more than twenty real designers are employed at Wycombe. Most of the manufacturers cannot afford a designer. In medium- and large-sized firms, the payment of the designer varies according to the size of the business rather than to the artistic quality of the work done. Most of the designers do not get more than £200 or £250 a year. But I was told that in three or four of the largest Wycombe firms salaries of £1000 a year were paid; and I myself met three designers who were being paid more than £1000 a year.

On the other hand, if one investigates the methods used for obtaining designs in those works where the best furniture is produced, one finds that they differ considerably from those outlined above. In one of the three firms to be discussed in this context, the proprietor himself was trained as a cabinet-maker and designer in a high-class furniture firm. For a long time he has done most of the designing himself, but now younger members of his family and a "school" of designers working in his style have followed. The influence exerted by him on the whole of the taste in modern English cabinet-making between 1905 and 1925 is obvious. In the second firm of extraordinarily high artistic merit it is also the founder and owner who is responsible for the style of the products. He started his business by learning the craft of cabinet-making and has made with his own hands many pieces of outstanding quality. He also designed most of the early work of the firm. Later on he was joined by another member of the family, a trained architect of progressive and uncompromising taste. To-day most of the creative work is in the hands of a staff designer—one of the rare members of the profession who combine high standards of art with technical and commercial knowledge. As the work of this firm was decidedly modern and unconventional from the beginning it proved difficult at first to find an adequate market. Some support seemed advisable, and this could be given with the aid of another line of business, in the hands of the same company. However, these initial difficulties were soon overcome, and a steady growth set in. This was considerably accelerated when the firm started the production of wireless cases for an enterprising radio firm. To-day a satisfactory balance is attained between mass production and highest quality cabinet-making. In our connection it is important to draw attention to this instance of the combination of some mass production "bread and butter" lines with the making of a comparatively small quantity of articles too advanced and uncompromising in design to be a sufficiently safe proposition by themselves.

In the third firm where furniture of the very best aesthetic standard was produced, a similar arrangement had been made.[1] Here the bulk of the output consists of well though conventionally designed suites at about £25–£30 retail price. But apart from these, this firm, about four years ago, started the manufacture of a special type of simple and pleasant modern easy chair and also an equally modern and simple type of unit or sectional furniture. In proportion and detail both the single units and the complete wardrobes, chests of drawers, etc., are equally perfect. The prices are high, owing partly to difficulties in advertising furniture under a brand name. The designs are derived from those created by the German architect Schuster in 1930. His patent on the constructional system was taken over by a distinguished architect in England who had to alter several of the sizes and proportions. The manufacturer to whom we have referred then undertook the production, and probably he, too, could not have afforded to take this step, unless he had been sure of the safe background given by his main business which constituted about 95 per cent. of his output, as compared with only 5 per cent. in the modern range.

Having enlarged upon these cases of furniture of an outstanding artistic quality, it must now be said that they are exceptional. On the whole, the taste in contemporary English cabinet-making is very unsatisfactory. The reason here is not

[1] In 1936 the firm in question gave up the production of the two modern lines described above. They are now carried on by a separate company, Plan Ltd.

so much the survival of period imitation, as we found in artistic metalwork, but rather the craze for the "modernistic" which we have described before when dealing with door furniture, etc. The change from traditional to "modern" forms took place in commercial cabinet production about six years ago. Now, I was told in a London departmental store, period furniture amounts to only 5 per cent. of the total turnover of the department. A similar figure was quoted to me in a departmental store in the North of England. In another, more expensive, store selling to a rather conservative class of Londoners (Kensington), period articles are bought mostly by the older generation, though in dining-room suites they still represent 75 per cent. of the sales. Bedroom suites and suites of settees and easy chairs are almost exclusively modern. But buyers in several cases seemed to have noticed a growing demand for reproduction work.* Practically all cheap mass-produced furniture is modern, or rather "modernistic". One cannot help regretting this deeply. The Morris Movement had resulted in a revival of some types of simple furniture of an almost perfect, and at any rate undated, character (Morris chairs, wooden bedsteads). Much of this survived the Edwardian Baroque, and can still be found, together with old-fashioned cretonnes and floral wallpapers, in certain private hotels and boarding houses in cathedral towns or the countryside. In the small new suburban houses round London, however, it seems completely superseded by the bastard deformities of the Modern Movement.

This situation must appear even more disheartening if the fact is added that many of the shopkeepers realize the bad taste of their cheap goods. In both the stores to which I went it was pointed out to me that changes for the better were noticeable only in the higher price range. The sole technical reason for this difference between cheap and expensive products is the fact that plain surfaces require better timber and more care than surfaces to be covered with stuck-on ornament. But that is of course an insufficient reason. Good shapes and good proportions certainly do not cost more than bad. Why then are they not to be found in cheap suites? In some cases it is obviously the fault of the manufacturer. If, as I found more than once, he calls good modern design "cigar boxes" or the "Assyrian craze", not much can be expected. But more often I found manufacturers convinced of the deficiencies of their own products. They tell you, however, that the public would not buy suites of better design. Not that they deny the trend towards simplicity. Instances were cited to me where some angles had to be straightened in order to make a piece more popular. But this process of simplifying must not, they say, be carried too far. Mirrors must remain oval, bedsteads sumptuously carved and moulded, wardrobes provided with some flowing fillets. Nothing as severe as the products of Central Europe would sell to the English public. The manufacturers are convinced of this, though they admit themselves that they cannot possibly know the exact opinion of the public. The case is here the same as in most other trades. Goods are sold to the buyers of furniture stores who may be true judges of the popular taste, or may be totally wrong, and thus deprive the public of their chance of seeing certain things and perhaps taking to them. And sometimes it must be questioned whether the manager of a shop refuses well-designed cheap furniture solely because he is so certain that the less wealthy class would not buy it. One case in point, and an instructive one, is this: In a high-class store at Birmingham, selling furniture, textiles and objects of industrial art, the proprietors refused to take any interest in my research, saying that if I succeeded in improving the taste cheaper shops everywhere would take up lines similar to the ones which so far they alone displayed, and this would be detrimental to their business. I found the same attitude in a large furnishing house in London, where it was explained to me that if the existing department for good and mostly expensive modern things were not so different from, and superior to, the bulk of their own goods, their "market would be upset".

Here we are faced, not with the conviction that the majority of the people would not buy the better design, but with the highly objectionable tendency to prevent people from getting it, in order to keep it exclusive to one class.

So much for the store. As for the producer, at least in one case of a mass-production factory, one of the managing directors said with some pride that their "production is so overwhelmingly extensive" that they are able "to create the popular taste".

This being so, it can hardly be denied that less blame for the shocking taste in most furniture is due to the public than is often supposed. Two arguments may be added to prove this. One is the success of the cheap furniture in beech and bastard mahogany produced by one of the firms visited. The commercial success of these satisfactorily designed products shows that the public—at least under certain conditions—buys simple and straightforward furniture without hesitation. The other argument is my questionnaire at the Midland Industrial Art Exhibition. It showed more unanimity than in any other instance in the welcome given to the Gordon Russell and Plan Furniture. Out of 71 answers, 19 contained some Gordon Russell products amongst the ones liked most, 14 some of the Plan exhibits; 13 visitors were in favour of some simple furniture made by another firm. This result appears even more remarkable if we consider the negative answers also. The pieces most universally condemned were a group of cane chairs, certainly well and competently made, but lacking that appearance of contemporaneity which characterizes the Plan and Russell products. In spite of the limitations of the questionnaire, one can hardly shut one's eyes to the fact that here a section of the public has distinctly expressed its predilection for the best type of modern furniture.

How then can the manufacturer insist upon saying that the English public would never buy unadorned suites? Has not the same English public bought the simple and pleasant chairs, tables and bedsteads which became popular through Morris? If these were not too plain, why should modern chairs or wardrobes be too severe for the same public? Or has the English public in general not taken to contemporary simplicity because it has never had a fair chance of seeing it? Could the manufacturer create a demand for something better if he really wanted to?

Another argument which in some ways can be regarded as telling in favour of such a view is the *history of tubular steel furniture*. Since the facts about the invention and the early stages in the development of tubular chairs are almost unknown, a few data may be given.[1] The first piece of steel furniture was patented by an American in 1904. It was a divan constructed not from steel tubes but from steel rods. About 1924, another patent was taken out in the United States[2] referring to a chair without separate back legs which was also made from steel rods.

[1] I am able to base the following account on detailed letters received from W. Gropius (Sept. 22, 1934), Marcel Breuer (Oct. 3, 1934), Miës van der Rohe (May 7, 1935) and A. Lorenz (June 12, 1935), and gladly take this opportunity of thanking them for their kind help.

[2] Mr H. E. Nolan.

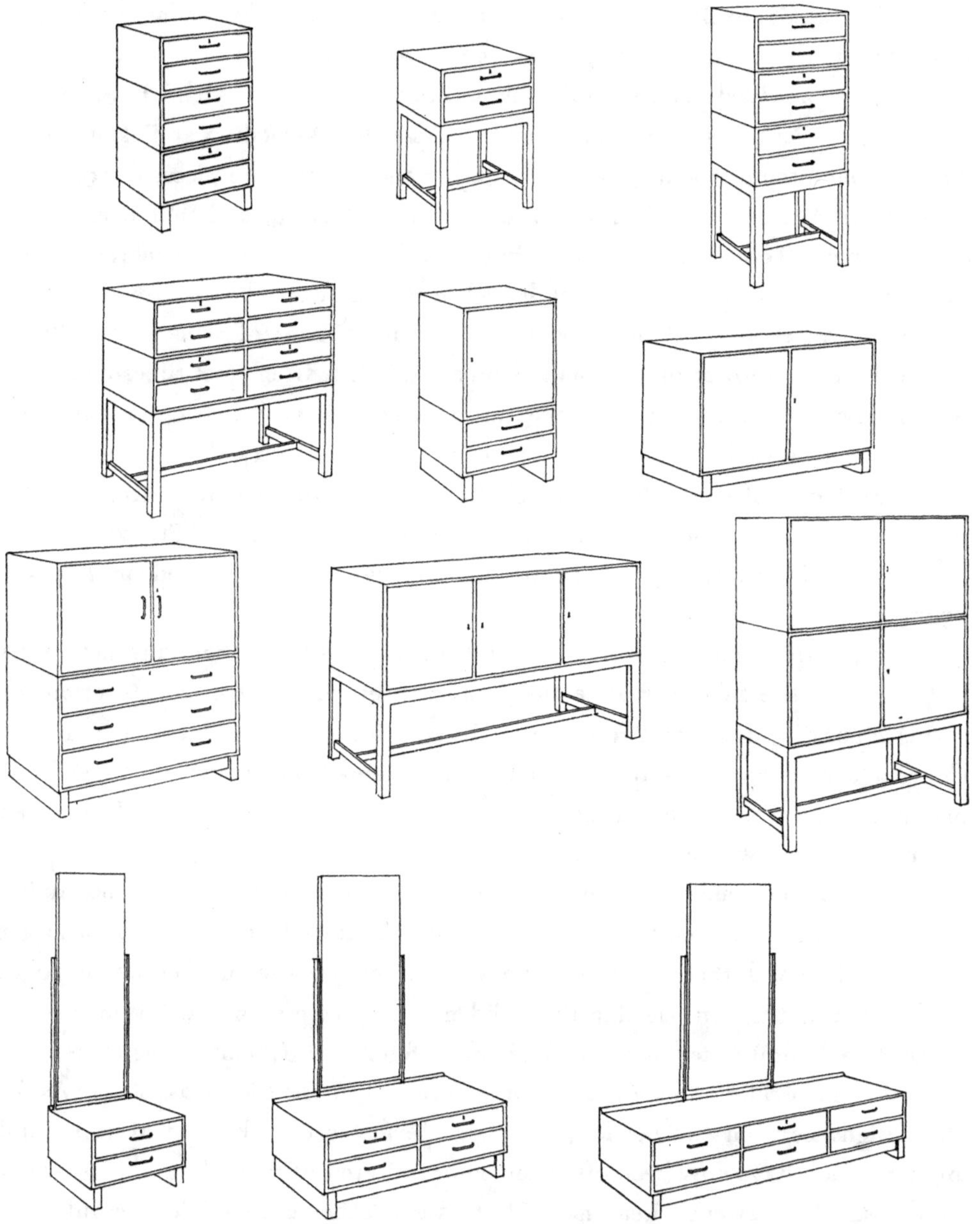

VI. *"Plan" furniture, designed by F. Schuster and S. Chermayeff and made by Plan Ltd. Various combinations of units.*

The first chairs using steel tubes were created by Marcel Breuer, who was then in charge of the cabinet-making class at the Bauhaus. That was in 1925.[1] In 1925–6, he provided the new Bauhaus buildings at Dessau with his furniture. The designs were carried out at a small locksmith's workshop at Dessau. Production amounted to about 500 pieces in 1926 and 1000–1500 pieces in 1927. In 1928, Herr A. Lorenz in Berlin took over the production, and thus became the first real manufacturer to turn out tubular steel furniture on a commercial scale. In the meantime (1926–7) the Dutch architect Mart Stam had created chairs of a new shape, entirely abandoning the traditional forms of wooden furniture. These chairs were first built from gas pipes and then from steel tubes and were without back legs. Thus their appearance came very near to that of the most popular models to-day. But in construction they still did not take advantage of the resilience of steel. In Stam's chairs the tubes had an iron core, in order to make them more solid. They were shown at the famous Stuttgart exhibition "Die Wohnung" (Weissenhof-Siedlung) which was opened on July 24, 1927.

Ludwig Miës van der Rohe, the last principal of the Bauhaus before its collapse, was the first to realize the great possibilities of steel construction for giving chairs lightness and resilience. His solution of the problem, which was the conclusive step in the development, was patented on August 24, 1927. The manufacturing was for the first few years done in the workshops of a firm called Berliner Metallgewerbe Jos. Müller.

Since 1929, Gebrüder Thonet, the famous mass producers of bentwood chairs, who some time later took over the Lorenz firm, started production on a larger scale. In England the first steel chairs were put on the market about 1930. Unfortunately the firm making them did not wish to give me any information. Since then several other manufacturers have followed. I went to see three, and enquired about sales in one departmental store and in one high-class shop. As for the designs now on the market, the most perfect are still those of Rohe and Breuer which have now been often copied or imitated with slight and artistically unnecessary alterations. Leaving aside this and other cases of designs fully developed in Germany before they reached England, the invention of new shapes

[1] Cp. *Das Werk*, vol. vii, p. 210, 1926.

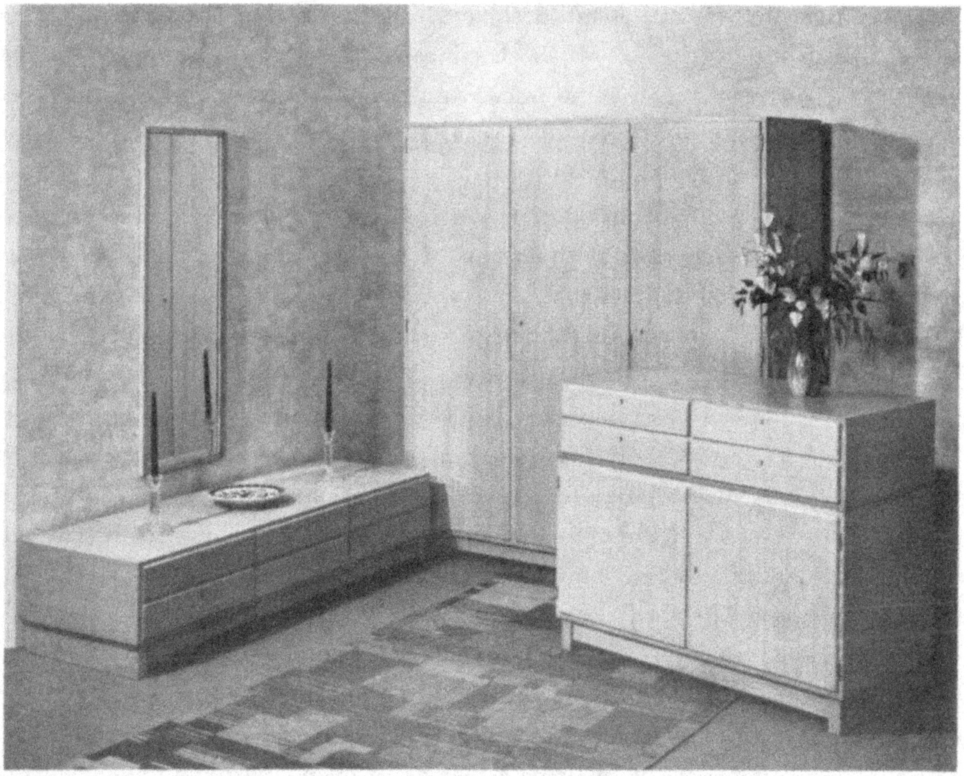

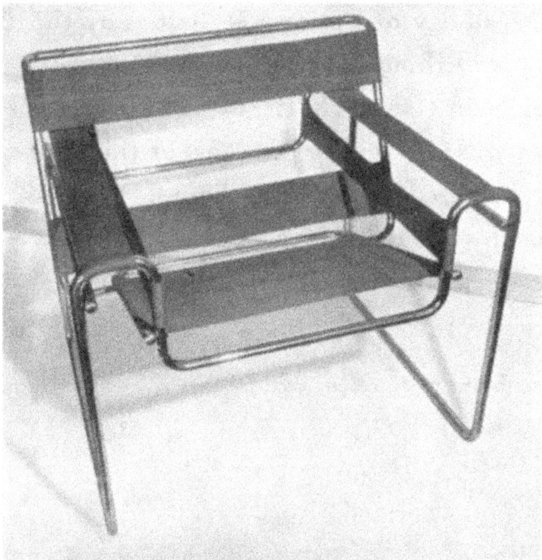

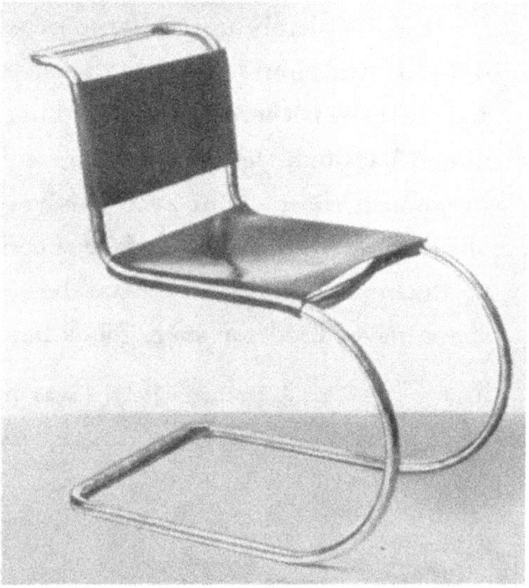

VII. *Some pieces of "Plan" furniture.*
The first tubular steel chair ever made. Designed by Marcel Breuer in 1925.
L. Miës van der Rohe's first tubular steel chair without back legs. Designed in 1927.

was in one firm the work of a staff designer, in the two other works it was done by the manufacturers themselves, both educated men keenly interested in the matter. One of them said that he finds it an easy and pleasant task to try out new forms, the other preferred to communicate with one of England's most advanced architects.

As far as the quantity of production is concerned, one firm mentioned 11,000 chairs in 1933; another several thousands of one very popular model in 1934; two works emphasized the fact that there was twice or even three times as much demand for tubular furniture in 1934 as in 1933.[1] The reason generally given for the rapid development of tubular furniture is fashion. That is unquestionably true. Certain shops or road houses will insist on steel chairs and tables because they regard them as smart. In my questionnaire the characteristic expression of this situation was that tubular furniture was mentioned in a remarkably high number of answers, 25 visitors putting one or more articles among the best liked, 17 among the least liked. This shows that the matter is, at any rate, controversial, and arouses strong interest in certain circles.

However, such an interest could not have arisen nor be kept alive, had producers not decided to push tubular furniture. This is important in a much wider sense. Here is a proof that manufacturers, directly they want to push a certain type of goods, can succeed in creating a fashion or even creating a true change of popular taste, which may be for the better or the worse. I do not deny that this is not a completely unassailable proof. The validity of the case is limited by the fact that steel chairs are still expensive (the type without back legs costs about 50*s*.) and their use is therefore confined to the wealthier classes. In the department store which I visited, the turnover in steel chairs was still only 5 per cent. of the sales in wooden chairs. But even this may be considered a surprising figure. In spite of all these limitations the fact remains that by the determined attitude of a few producers a type of goods has been made saleable, although of a material and shape never used for such goods before.*

[1] Unfortunately, I was unable to obtain more recent figures

TEXTILES: I. FURNISHING (AND DRESS) MATERIALS

The designer for industrial art has to face two fundamentally different problems. His task can be to shape objects or to decorate them. Designing a gas fire is almost entirely a matter of shaping it, designing an electric fitting or a wardrobe may mean decorating surfaces as much as shaping volumes. The creative process in designing a printed cretonne or linen or chintz is exclusively the invention of a pattern for application to a material of hardly any intrinsic aesthetic significance.

However, the designer for the textile trades is not concerned with printed fabrics only. In the creation of a new texture for a tweed, the artist is, generally speaking, in the same position as in the creation of an electric pendant or a piece of furniture. No longer has he to consider a flat surface only; he has now to think three-dimensionally, in terms of structure, of thickness and thinness, roughness and smoothness, etc. Hence, in dealing with furnishing fabrics, an explicit distinction has to be made between "woven" and "printed" textiles. The designer who works for "prints" does not require much special technical knowledge, beyond some familiarity with questions of repeat, and with the advantages of handblock printing, screen printing and roller printing. It is a very significant fact that almost all the great personalities from outside the trade who have influenced textile design within the last fifty years are known for their printed and not their woven fabrics. Voysey's designs for factory production, the designs of Roger Fry's Omega Workshops between 1912 and 1918, and the designs of Duncan Grant, Paul Nash, Marion Dorn were and are exclusively or predominantly done for the printers.

To invent new effects in weaves is impossible without a live knowledge of weaving.[1] Whether this should include personal experience of all the processes of machine production, or whether some basic facts which any hand-loom can convey are all the designer should be acquainted with, I do not feel entitled to decide. Two manufacturers whom I met emphatically stressed the similarity of

[1] Cf. *Design and the Cotton Industry*. Board of Education, Educational Pamphlet No. 75, pp. 5–6, 1929.

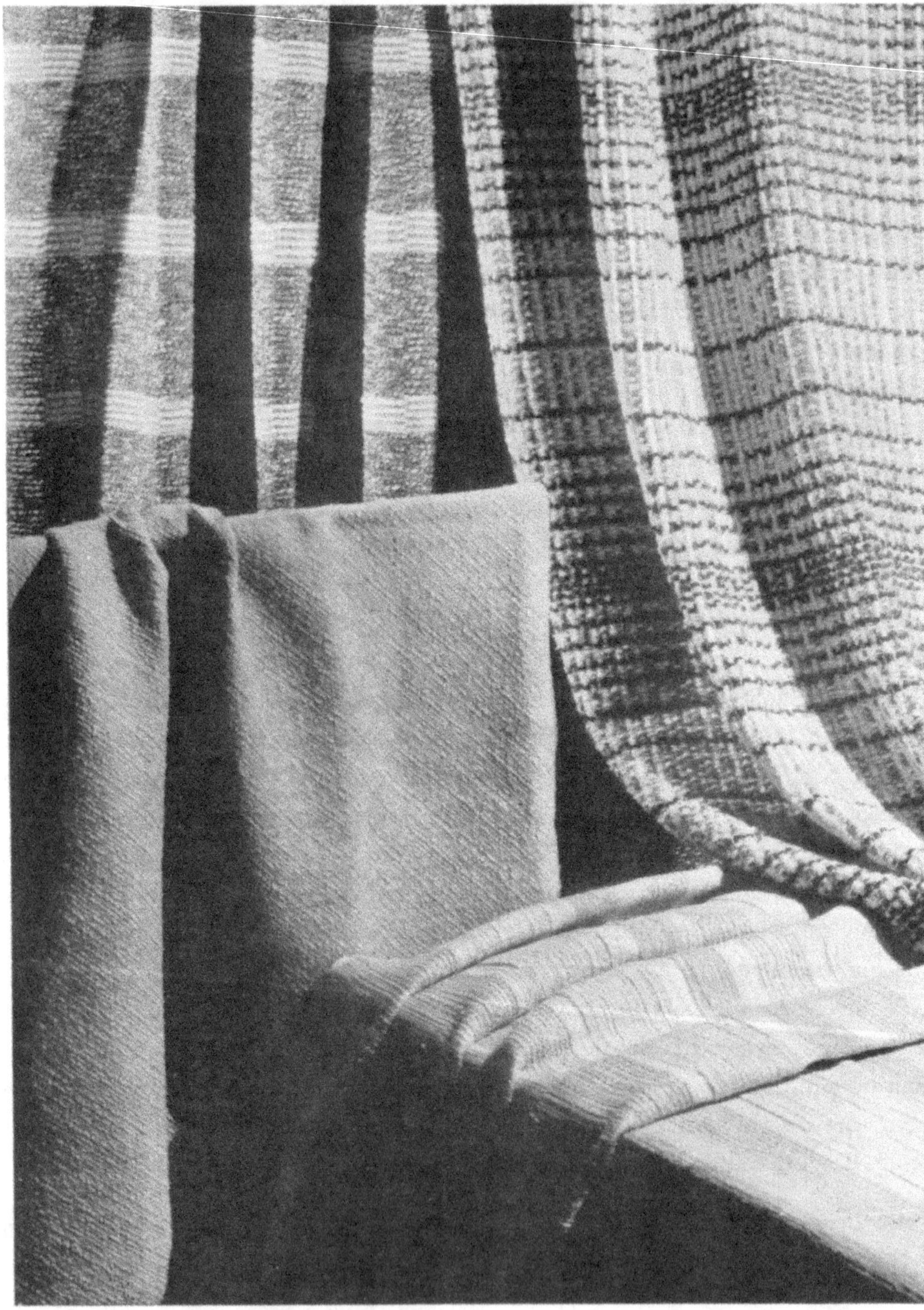

VIII. *Some furnishing materials by Donald Brothers ("Old Glamis"). Top left "Moidart", top right "Duntulm", bottom left "Roslyn", bottom right "Leven".*

hand-loom and Jacquard loom.[1] English art schools seem to be of the same opinion; otherwise their weaving equipment could not be so poor. Power looms are as a rule only to be found in technical colleges. Thus, while designers for printed materials may come from art schools, designers for woven fabrics usually acquire the bulk of their knowledge in the drawing office of a mill and take some additional afternoon or evening courses at a technical school. Many manufacturers regard this as the best working education and speak of art schools in a rather disparaging way. Among their arguments against a higher form of artistic training for designers one above all is worth mentioning. Several times doubts were expressed to me as to whether it would pay to create ideal schools which could not help turning out far more first-rate designers than could be absorbed. One good artist can easily produce more patterns in a year than even a big firm can put on the market.

Nevertheless the textile trade is, if not the largest, one of the largest users of designs in Britain. I interviewed eight manufacturers, one merchant, three designers, and the textile buyers in six department stores. Of the manufacturing firms one was a house famous for dress materials, employing a staff of over 4000, one a well-known mass producer of furnishing fabrics (over 1200 employees), one a good Manchester firm producing mainly medium-price goods (about 700 employees), one a factory renowned above all for their reproduction weaves and prints (550 employees), one the most adventurous producer of furnishing materials in Britain, one an equally famous factory the reputation of which is based predominantly on their crashes and linens (250–300 employees), one specializing in prints which are again mainly reproduction work (300–400 employees), and one a large mill where spinning is more important than weaving and dress materials more important than furnishing materials (2000 employees). Among the designers whom I interviewed one is the head of the works studio of one of the best English firms, one runs a studio in Manchester, the third a studio in London. As to retailers, I received information from the textile departments of four of the best known London stores, of which one is comparatively low in price, one the highest quality department store in Britain, the third is in Oxford Street, the

[1] Cf. F. E. Donald, Address given to the Red Rose Guild and the Design and Industries Association, Manchester, 1935.

fourth in the south-west district of London. The fifth store is in the Midlands, the sixth in Lancashire.

To give an idea of the quantities of designs required for the market, the following figures may be quoted: the mass producer of dress materials: over 800 a year (of which incidentally about 200 are for ties, about 100 for shirting, and about 50 for handkerchiefs); the mass producer of furnishing materials: about 50 new prints, and about 100 new weaves; the other firms interviewed: 70 prints plus 50 weaves, 80–100 altogether, 30–50 altogether, etc.

Apart from the manufacturers there are also merchants who order materials to be printed to designs supplied by them (the merchant whom I saw stated that 70 per cent. of the patterns that he runs are his own designs), and printers who offer to provide the designs when they are asked by mills to print cloth for them.

Designs for furnishing materials are as a rule produced by mill studios (and sometimes London studios of Lancashire, Yorkshire or Scottish mills) or by independent commercial studios belonging to so-called "public designers". Free-lance artists are numerically negligible though artistically important. Capt. W. Turnbull in an instructive article on textile design[1] states that 42 per cent. of designs come from works' studios, 42 per cent. from foreign commercial studios, 13 per cent. from English public designers, and 3 per cent. from free-lance artists. My investigations on the whole confirmed these figures: the large firm producing dress materials, 25 per cent. studio, 75 per cent. bought from outside; one of the manufacturers of printed fabrics, 20 per cent. and 80 per cent.; the mass producer of furnishing fabrics, 66 per cent. and 34 per cent. in woven fabrics, 70 per cent. and 30 per cent. in prints. The three most progressive manufacturers of furnishing materials gave me the following figures as their ratio of studio-produced to purchased designs: (1) some years 50 : 50, some years 60 : 40; (2) in prints 40 : 60, in weaves 95 : 5; (3) in prints majority bought, in weaves vast majority worked out between director and manager.

In the last two cases the contrast between the situation in designing for weaves and for prints is striking. Only men who can keep in close contact with the technicians and workers are suitable for the creation of new weaves; the outside

[1] *The Journal of Careers*, vol. xiv, pp. 91 *et seq.*, 1935.

artist who is supposed to be more imaginative or more facile, because he is not restricted by the narrow life of a mill, is more appropriately employed for printed patterns. However, one has to beware of sweeping generalizations. A few of the best manufacturers give their designers considerable freedom, send them on journeys abroad, allow them to arrange their time as they like, and pay them generously.

The average payment of the head designer in a firm of some standing is about £400–£500, or anything up to £1000 and more.[1] Studios have permanent staffs of five to twenty men. But a great deal of their time is spent in working out colour ranges and in finishing sketches bought from outside designers. Thus the mill designer does not always find as much leisure for creative work as he would like to have and sometimes not as much as the manufacturer would like him to have in order to be less dependent on outside sources.

Of the commercial designers, a good many in England and on the Continent run large studios and employ 30 or more designers, finishers and apprentices (the proportion between these being, according to the Board of Education Report mentioned: 35, 55, 10 per cent.). The Manchester studio that I visited produces over 100 designs a year, most of which are for printing and are sold finished to roller size. The fees which commercial designers receive vary according to the use to which the design will be put and the degree of finish. The Board of Education pamphlet gives as an average price for a design for dress materials £4, and for a furnishing fabric £12–£15. The two mass producers of dress materials whom I asked about this said: £3–£20, and for a sketch as a rule not more than £1–£2. For furnishing fabrics I was given the following figures: £7–£10 and up to £15; £7–£10 and also down to £3 and up to £15; £5–£15; £8–£10.

Besides English commercial designers, French and German studios are of foremost importance. There is hardly any difference in price between British and Continental sketches, though those from France and Germany may be slightly cheaper. The proportion between English and foreign designs purchased by firms

[1] Cf. "Design and the Cotton Industry", *l.c.*: mills with 200–400 looms £200–£250; larger mills up to £400 and considerably more. Cf. also Capt. Turnbull, *l.c.*: £350–£500 and up to £1000. Capt. R. D. Simpson in *Journal of Careers*, vol. xiv, p. 89, 1935, writes that salaries range between £500 and £1500. Of the industrialists whom I interviewed one said: rarely under £500; another, £750–£1500.

varies greatly according to the artistic and political views of the manufacturers. The following are the percentages of foreign designs used in the firms which I saw: 5, 5–10, 35, 65 per cent. (almost all French—the figure comes from one of the producers of dress fabrics), almost 100 per cent. in prints and about 85 per cent. in weaves (of which two-thirds are German).

The highest fees are naturally paid to some free-lance artists of renown, painters and designers whose names carry enough weight to justify an initial expense of £18 and more before the production of the new pattern is started. For such is the usual fee in these cases. It often goes up to £20 and more and only rarely drops below £10. But there are only very few artists who are paid such fees. The majority of free-lance textile artists (designers as well as independent handweavers) can hardly make a living out of their work, although very often the introduction of new tendencies into the trade is due to these "unpaid research workers" (Turnbull).

Provided with the above data, we can now pass on to some questions connected with the artistic situation in the textile trade. The Modern Movement (including "Modernism") in England, in spite of some early attempts such as those of the Omega Workshops, and in spite of the Paris Exhibition of 1925, had hardly any influence on the trade before about 1930. An exhibition at Waring's which Serge Chermayeff arranged in 1928, and to which we shall have to refer later, stands at the beginning. The new activity of the Edinburgh Weavers in a strictly modern spirit was the second symptom, the beginnings of Allan Walton's printed satins with modern, mostly abstract patterns followed soon. To-day period reproduction has become rare in woven fabrics, except for certain damasks, brocatelles, velvets, etc. Again some of the figures which I was given may serve to corroborate this. Three manufacturers: 97½ per cent. modern, 50 per cent. modern, 85 per cent. modern. Commercial designer in London: 75 per cent. modern. Department store in Oxford Street: 90 per cent. modern; department store in the south-west district of London: 90 per cent. modern; higher price store in the North of England: "almost all" modern; big department store in the North: "mostly modern"; but, characteristically enough, highest class department store in England: 60 per cent. modern, 40 per cent. period, and a marked upward tendency in period materials.

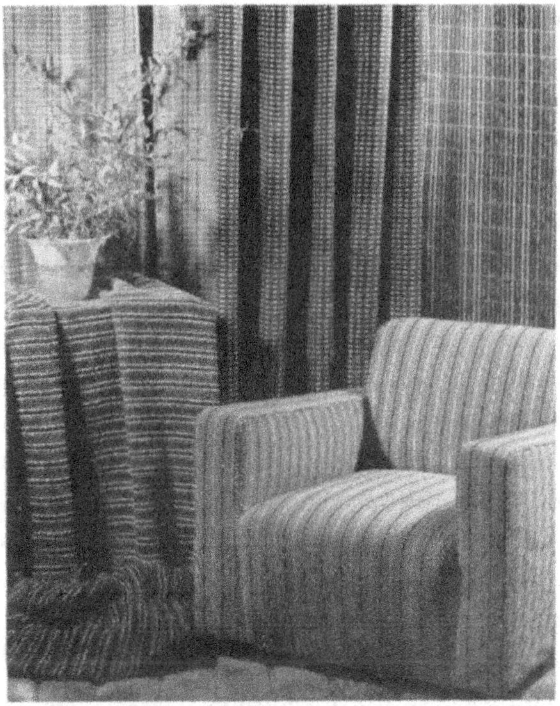

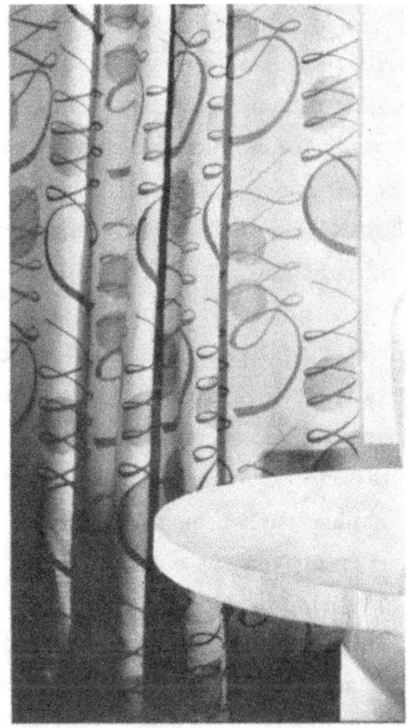

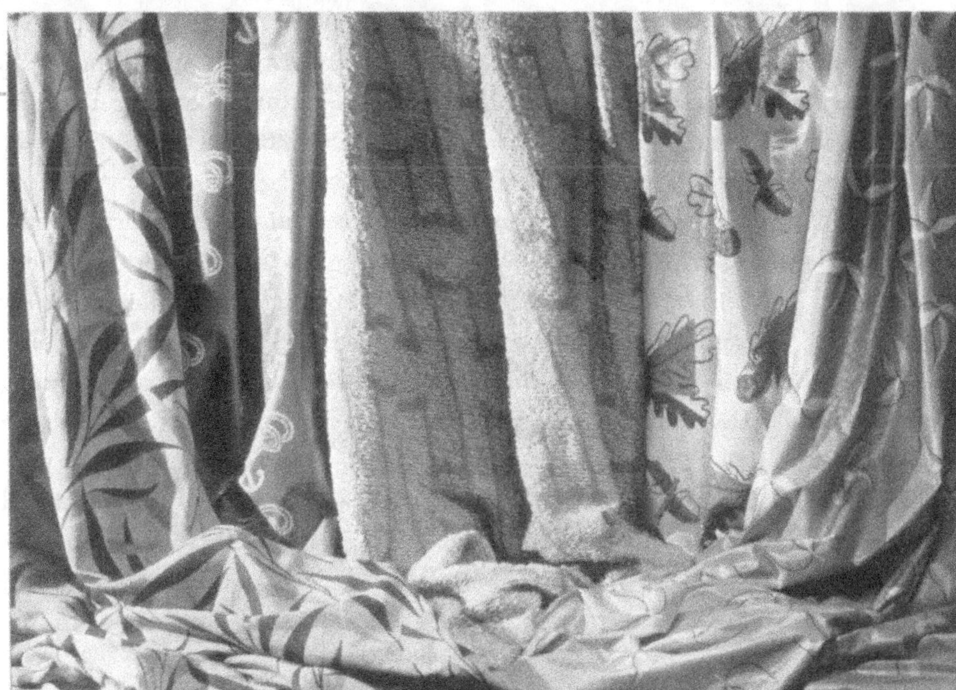

IX. *Welsh tweeds woven for Gordon Russell Ltd.*

"Chorale", printed linen designed by Marion Dorn (Old Bleach Linen).

Furnishing fabrics by Warner & Sons: From left to right: "Bamboo Grass" (designed by H. Woodman), "Anchors" (designed by Marion Dorn), "Waltham" (designed by R. Silver), "Acorn and Oakleaf" (designed by Marion Dorn), "Falling Leaf" (designed by Marion Dorn).

In printed fabrics the situation is entirely different. Here period imitation is still prevalent everywhere, and at the moment perhaps stronger than about eighteen months or two years ago. Period prints, in the same firms of which the figures for modern weaves were quoted just now, come to: 10 per cent. only; 75 per cent.; 30 per cent. pure reproduction plus a large percentage of "period floral" in a more general sense; 75 per cent.; majority; large majority; 95 per cent. in linens and 90 per cent. in chintzes; 35 per cent., but increasing; "almost all".

It is not easy to understand the reasons for this contrast between woven and printed. The most likely reason seems to be the extraordinary strength of tradition in English printed floral decoration. For such a long time English patterns had served as examples for the whole Continent that modernism proved powerless. Moreover the good designs of this type possess such an exquisite balance, such a perfect blend of realism and pattern effect, that no need was felt for improvement or change. Another argument, of a technical nature, may also have some bearing. The preparation and reproduction of a new design for weaving is far less costly than for large-scale printing. The cutting of the cards for a Jacquard loom may cost anything from £10 to £50. After selling 2000 or 2500 yards or, in some cases, less than 1000 yards, this initial expense is paid off. Moreover a modern design does not require as a rule so many cards as a period damask. The position is different in machine printing. The engraving of the rollers costs between £50 and £300. The total expenditure, including trials and samples, before starting to market a new pattern, may be anything from £250 to £600. Consequently any design of which less than 3000 or even 6000 and 10,000 yards are sold, must be considered a commercial failure. The corresponding figure in the Board of Education Report, referring to the cheaper Manchester trade only, is 30,000 to 50,000 yards. No material is called a best-seller unless over 60,000 yards are sold. All this, of course, does not hold good in screen printing or handblock printing. The total initial cost here would scarcely exceed £50. So here for once is a legitimate reason why in a trade the most progressive designs are sold at the highest price.

In weaving no such reason exists. Although certain qualities of feel and weight may be dependent upon expensive yarns and expensive methods of production, there is no justification for the cheap and nasty effects of some cheap upholstery and curtain fabrics. None, that is to say, unless those manufacturers are right who, in the textile industry just as in the industries discussed in the previous sections, assure you that it is the taste of the public that demands bad patterns. In fact, one of Britain's biggest textile manufacturers, a young, eager and modern-minded man, had to admit that amongst his best-selling woven materials there were two of a thoroughly bad and vulgar type. However, two is no great matter; and as my illustration shows (p. 54), most of the best sellers are at least unobtrusive and some are quite pleasing in pattern. One can confirm this by going along Oxford Street, New Oxford Street and Tottenham Court Road and looking at the windows of the furniture stores. One may dislike the heavy effects of big waves and chevrons on the "suites" of settees and easy chairs, or the cheap sheen and jagged patterns of cotton and artificial silk curtains, but atrocities such as those which shock us in the furniture shown in the same windows are extremely rare.

And on the other hand, woven fabrics of the highest artistic standards, standards not surpassed anywhere, are not at all rare. British tweeds, for dress (Munro, Gardiner) as well as for upholstery, the linens and crashes of Donald's, the interesting and sometimes enchantingly clever weaves of the Edinburgh Weavers, to mention only a few outstanding examples, would hold their own in Germany, in Sweden, and in France.

Generally speaking, it seems to me that in the ordinary English commercial production, simple texture effects or slight patterns such as indistinct stripes, etc., are aesthetically more successful than the more explicit motives. Daringly new decorative effects are risked by only a very few manufacturers, and do not always succeed so well as in some countries of a lighter and more playful disposition. Nor are strong and pure colours the forte of British commercial production. It is extremely difficult for the retailer or the public to get certain clear yellows, blues or greens without having recourse to foreign materials. However, I do not know of any country where some such criticism would not be equally justified. And as—to repeat this—no trade in any country produces, under present social conditions, more than 15 or at the most 20 per cent. of its wares in creditable taste, the state of affairs in English woven fabrics is undeniably gratifying.

The artistic value of English printed textiles is no lower. I know that, in

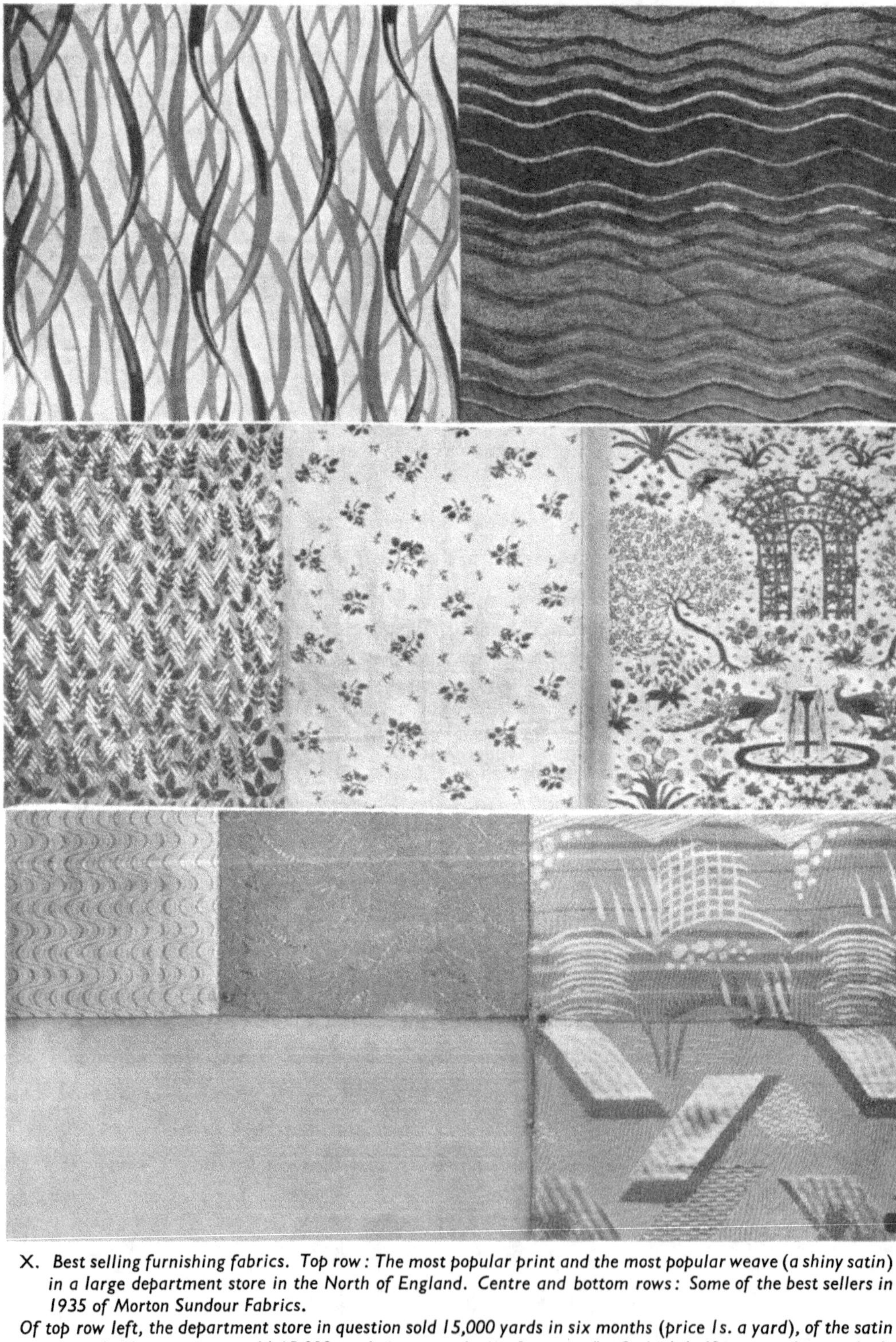

X. *Best selling furnishing fabrics. Top row: The most popular print and the most popular weave (a shiny satin) in a large department store in the North of England. Centre and bottom rows: Some of the best sellers in 1935 of Morton Sundour Fabrics.*

Of top row left, the department store in question sold 15,000 yards in six months (price 1s. a yard), of the satin top right the same store sold 65,000 yards in a year (price 2s. a yard) of which half in one colour, a shade of honey or buff.

making this statement, I shall be contradicted by a good many passionate supporters of the modern movement who would like to destroy every bit of reproduction work. I must confess that in textiles I cannot feel like that. As I have said before, it seems to me that the best English prints have achieved a perfection which is undated. Of course, here again these are the exceptions. There are a good many unsatisfactory cheap materials. On the other hand firms such as Warner's, Turnbull & Stockdale, and others produce printed linens and cretonnes with period floral patterns of exemplary design which sell at about 3*s*.–4*s*. per yard.

Good English prints of contemporary design are still rather unusual, less usual now perhaps than they were in the days of Voysey. What there are have all been created within the last three or four years. It is the merit of Allan Walton to have seen this gap in English textile art and to have run the risk of issuing fabrics designed by outstanding young painters. Of late his initiative has been followed by several other firms. Amongst designers Marion Dorn should be especially mentioned. Various excellent foreign designers have also been employed by British firms. Moreover there are in the studios of the leading factories some artists of the highest standing.

The social and economic conditions and the artistic standards of British textile design have been outlined. It now remains to seek for reasons that might explain the exemplary quality of some, and the creditable quality of most, British furnishing materials. A few such reasons have already been discussed. Of prime importance no doubt is the strong and indigenous tradition in prints as well as in weaves. English prints have flourished ever since the days of the Tudors and right down to the time of Morris and his followers; and British woven tweeds, etc., of perfect design and workmanship were produced by the weavers of Scotland and Wales even after the Industrial Revolution had destroyed almost all village crafts in Britain.

The initiators of the modern movement in the English textile industry, men such as the late Sir Frank Warner, Sir James Morton, the late D. T. and Mr F. J. Donald, consciously transferred this unconscious tradition of exquisite weaving to the factory and carefully developed it there. These same men induced the great designers of the late Victorian decades to work for commercial production. Their keen interest in the problems of industrial art is undoubted. It was this interest that led one of the younger manufacturers whom I met to go to Germany and Austria in 1927 in order to find out more about the new Central European style of decoration which had startled him at the Paris Exhibition. I found the same

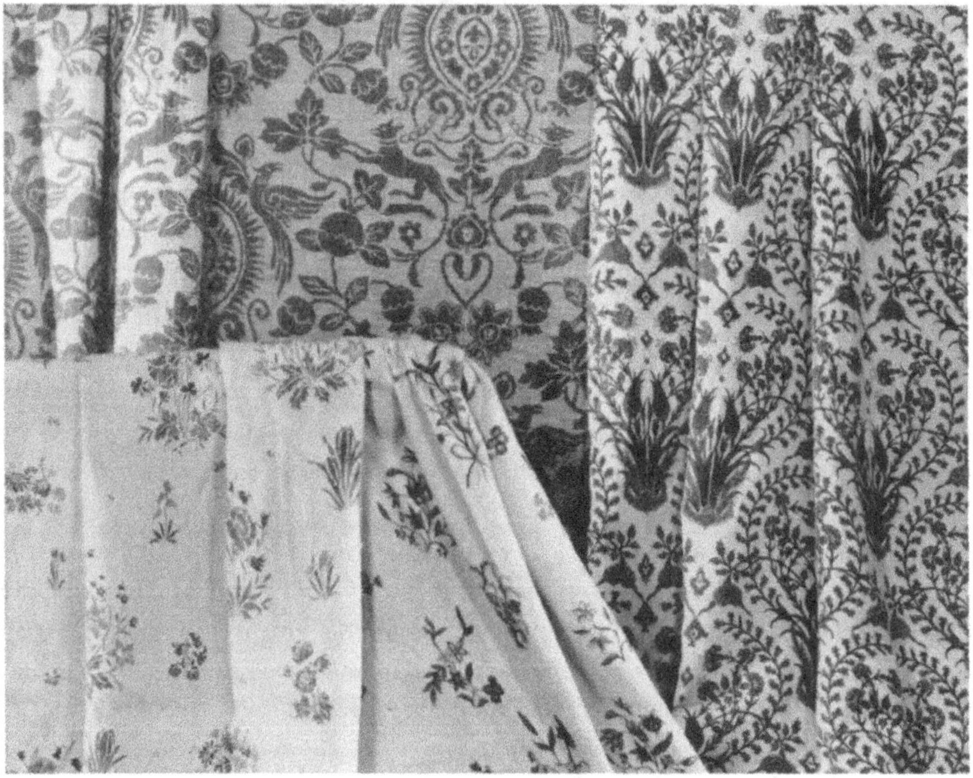

XI (*a*). *Printed furnishing materials.* (*Donald Bros.*)
Hanging: left "Palermo", right "Astrabad"; draped: left "Chandos", right "Pembroke".

interest in the intensely appreciative manner in which several other directors and managers explained to me new weaves and subtle new colour schemes, a manner very similar to that of obsessed picture collectors and connoisseurs. The same interest again induced Sir Frank Warner to work for the establishment of competitions for design at the Royal Society of Arts, and Sir James Morton to do what he did for the Edinburgh Weavers.

The history of the Edinburgh Weavers is of such paradigmatic importance in

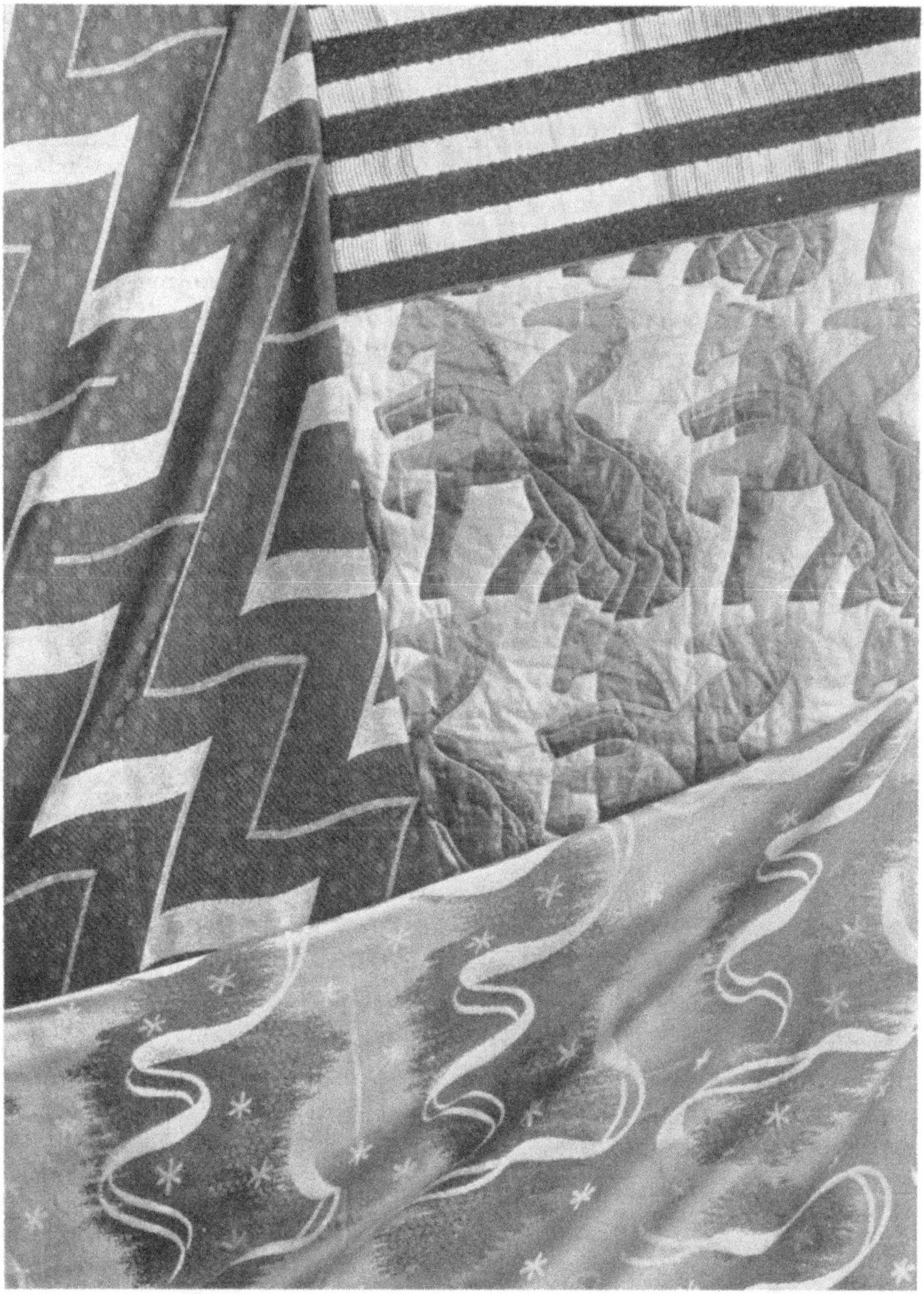

XI (*b*). *Some furnishing fabrics by Edinburgh Weavers: Left "Deveron", right top "Lindeth", right centre "Cavallo" (designed by H. Aufseeser), bottom "Milky Way" (designed by Ashley Havinden).*

our connection that it may be justifiable to relate it here in some detail. The firm was started as a handweaving shop (E. and A. Hunter) at Letchworth. It was then taken over by Dr Morton as an independent unit within the big mass-production combine of Morton Sundour Fabrics, with the explicit intention of creating a laboratory for the best modern textile art. Dr Morton was induced to risk this departure, the financial success of which was inevitably doubtful, by a strong feeling of responsibility for the value of his products, which is the highest ethical quality of the industrialist. With the aid of the profits from his "bread and butter lines", Dr Morton steered the Edinburgh Weavers through their first difficult years, and now they are well established and successful, beside being the most adventurous firm in the country.

To sum up: the two main results to be learnt from the data contained in this section are that a strong and live tradition can keep the products of an industry from the worst horrors of bad taste; and that the determined attitude of a group of manufacturers is bound, in the long run, to react upon the prevailing standards in a trade. It is essential to keep this in mind if we are to understand the striking contrast between the artistic situation in British furnishing fabrics and that existing in British carpets.

TEXTILES: II. CARPETS

My main problem here was this: Why are modern carpets in England so appallingly bad in design? Why is it almost impossible to find any machine-made carpets with an acceptable pattern and a tolerable colour scheme even in large specialized shops in London? To find an answer to these alarming questions it was important to have especially detailed evidence from carpet manufacturers.

I went to see the heads of eight carpet factories, of which five are amongst the largest in the country (with 3000, 2000, 2000, 1800 and 1400 employees respectively). Two of the other firms I visited were of medium size (with 700 and 350 employees), one was a small firm producing expensive Wilton carpets only. I completed the investigation by interviewing the heads of the carpet departments in three London stores.

The carpet trade is a comparatively small trade in England, both for size and number of firms. The *Carpet Annual* of 1935 enumerates 102 manufacturers; but only about 30 are of any importance. About half of these are at Kidderminster, a place of 30,000 inhabitants, which was once a pleasant country town specializing in hand-weaving, but is now a bleak and smoky agglomeration of mid-Victorian buildings. Other factories are in the West Riding of Yorkshire and in and around Glasgow. The whole of the trade gives employment to some 20,000 or 25,000 workers. This means that my information represents about half of the whole carpet trade. I saw six works at Kidderminster, two in Yorkshire, none in Scotland.

In spite of its moderate size, the carpet industry is certainly one of the biggest users of designs. The largest firms which I saw require about 200 to 300 new patterns each year, the less extensive ones 40, 60 or 70. These high figures are mainly due to the introduction of "Modernism" into the trade. Before and immediately after the war a good design would have had a run of six years or more; now some are dropped after a year or eighteen months. Well-selling lines are kept for two or three years; four or five are exceptional.

This large demand for designs is met by the same means as in the cotton and wool trades. Both industries find it profitable to use commercial studios apart from their staff designers. The "public designers" play only a slightly less important part than at Manchester. The ratio of staff-produced to bought carpet patterns varied in the firms which I questioned between 65 : 35 and 40 : 60. The staff in the designing offices of a carpet manufacturer is usually more numerous than that found in cotton mills. The five large firms mentioned each employed 25 to 60 draughtsmen and designers. In Kidderminster these frequently attend evening or even afternoon classes at the art school. In other works a complete training in the studio is considered more useful. The head designers, sometimes trained in the art schools, but more often grown up in the industry, may earn £350 in small firms, £450 or up to £600 in medium firms, £800 and more up to £1500 in the few large factories existing in the British Isles. Some of these have a separate designer in London who is supposed to have a better chance of keeping abreast of new fashions and Continental developments.

Of the commercial artists the majority are foreigners. Most of these do not specialize in carpets, but also do sketches for wallpapers, ties and dress materials, so that possibly I came across designs from the same studio in various industries.

These designers sell almost exclusively sketches, whereas the "public designers" in Kidderminster usually prepare finished drawings on point-paper suitable for immediate production. Prices vary between £2 and £10; £5 seems the average payment for a Continental sketch. As a rule more is paid for a finished drawing. Free-lance artists of renown get better fees; but carpet manufacturers of late have made very little use of painters or architects for improving the standard of their stock designs. They doubt whether such experiments could be a commercial success and are frightened of dealing with artists whom they call highly strung, touchy, inaccessible to commercial reason, and not sufficiently acquainted with the processes of machine weaving.

One manufacturer emphasized that the change from hand-loom to power-loom, from the hand-woven or hand-tufted to the Axminster or Wilton carpet, had brought such a change in basic premisses of design that now no art-school training in experimental hand-weaving could be of any use to the future designer of machine-made carpets. I cannot judge the truth of this statement. It seems strangely contradictory to those quoted above as to the fundamental similarity of designing for hand- and machine-made furnishing materials. The fact remains that the designer in a firm, as things are now, may never have done any weaving himself, nor have worked at a power-loom either in his art school or in the factory.

This separation between design and execution persists through the whole industry. At least 99 per cent. of the carpets and rugs produced in Great Britain are machine-made; in all these the worker has nothing at all to do with design. And even in hand-knotted wares it is only in exceptional cases that invention and craftsmanship are in the hands of the same person. The number of studio artists who weave or tuft rugs is negligible.

Amongst machine-made carpets, three types are particularly important: Wilton, Spool-Axminster (also called simply Axminster) and Chenille (also confusingly called Chenille Axminster or Patent Axminster). The weaving of Wilton carpets (and a more recent type of Axminster carpet, called Gripper Axminster) is based on the Jacquard loom as is that of most furnishing and dress materials. The main principle here is that long strips of strong paper (the Jacquard cards) are first perforated in such a way that each hole corresponds to one "shot" of the future carpet. The cards are then placed above the loom, and while they move along certain threads of yarn are, according to the arrangement of the holes in the cards moved, gripped by a kind of tongs, cut to the length required and inserted into the pattern.

The principle underlying the tufting of Axminster carpets is fundamentally different. Here the first stage is to prepare spools, on each of which is rolled side by side as much yarn of different colours as is needed for tufting one horizontal section of the design across the carpet as often as this particular section recurs in the "repeat" of "piece-goods", or in the quantity of individual carpets (with no "repeat") which the manufacturer wants to produce at a time. All these spools are then fixed between two chains and as the chains pass through the loom one spool after another arrives at a certain place where, for each horizontal row of the carpet, as much of each colour is cut off and tufted in as is necessary to form that part of the pattern. Then the next spool follows, and the first does not reappear until the same part of the pattern is repeated. The length of the chain thus depends on the size of the repeat or on the size of the carpet.

Chenille carpets are woven and not tufted. The weft consists of caterpillars of chenille which contain the whole of one row of the pattern and are made in a preliminary operation identical to that of chenille-weaving and chenille-cutting in general. The invention of this technique dates from 1839, and that of Spool Axminster carpets from the early 'seventies. The great advantage of Axminster is that an unlimited range of colours can be used in each section, whereas in Wilton and Gripper the number of colours is restricted.[1]

Wilton carpets are more expensive than Gripper, Chenille, and most Axminster carpets. This is due to the fact that in Wilton weaving a denser surface is produced than in most Axminster tufting. One may take 11 "shots" per linear inch as an average in Wilton, 7 or 8 in medium-priced Axminster, and 5 or 6 in cheap Gripper Axminster. Owing to this fact, the volume of trade in Axminster and Chenille carpets amounts to about four or five times as much (in yardage) as in Wilton. Cheaper even than Axminster are most hair carpets; more expensive even than most Wilton goods are machine-knotted (Machine Smyrna) and hand-knotted carpets.

[1] For illustrations of the three techniques, cf. *The Architectural Review*, vol. lxxix, p. 186, 1936.

The distinctions of price and technique correspond to distinctions in design. The main dividing line to be drawn in carpet design is that between Oriental and modern (or "modernistic"). Though there are a number of firms going in exclusively for the imitation of Persian and Chinese carpets, an imitation performed sometimes with remarkable skill, the bulk of the trade is now modern (or "modernistic"). It is hardly possible to give an exact ratio of the two kinds: 25 : 75 or 35 : 65 may not be far from the truth. Differences in the popularity of modern designs exist in various parts of London and Great Britain. Harrods show more Persian patterns for their Kensington clientele than do their cheaper competitors in Oxford Street for their customers. One manufacturer told me that Scotland is rather set against bright colours, another one that Oxford is against the modern trend altogether. The United States also seem to be remarkably faithful to the Oriental taste. The catalogue of a mass-production firm which, so I heard, controls more than 20 per cent. of the American market, contained fifteen Persian and only three modernistic patterns.*

On the whole it can be said that the higher the quality and price of a carpet the more traditional and reticent will be its pattern. The director of one firm said that although Oriental designs did not amount to more than 35 per cent. of the firm's total output of Axminster and Chenille carpets, in high grade Axminsters Oriental designs amounted to 80 per cent. The small manufacturer of expensive Wilton carpets whom I interviewed had not a single modern stock pattern. He pointed out to me that only cinemas and hotels would order expensive long-wearing carpets with "smart" patterns, and these, of course, are special jobs rather than stock products. The ordinary trade in high quality carpets is as a rule concerned with wealthy people of a certain conventional class who either prefer period houses, period furniture and Persian carpets, or else do not want to run the risk of buying something that may later look "dated".

In modern carpets there are a few exemplary designs to be found, created by such artists as Marian Pepler and Marion Dorn; but these are, apart from rare exceptions, all hand-tufted. The exceptions mentioned are machine-knotted goods, also of a fairly high price. Amongst plain carpets sold by the yard there are some obtainable in satisfactory shades. It is, however, not an easy task to pick these out from a host of vulgar colours, at least in most carpet stores. As to all the rest, one may safely say that directly one gets away from period imitation there is hardly a single piece of acceptable taste. This assertion can be confirmed by looking at trade magazines (and the competitions appearing therein), trade catalogues and the contemporary plates in Mr C. E. C. Tattershall's book, *A History of British Carpets* (London: Lewis, 1934).

The dominant patterns for the last eight or nine years have been derived from a deplorably misunderstood Continental cubism, the prevailing colours being a brown, a blatant orange and, in more recent years, a grass-green no less blatant. The introduction of some shades of pink and strawberry is an enrichment developed during the last two or three years. Some signs of improvement can be seen in a more frequent use of light and dark grey; but these are faint signs.

On several occasions I asked manufacturers why they did not change the style of their carpets. One of their answers was that changes, unless commercially necessary, are far too expensive to be risked. It would be unjust not to quote the figures which they put before me. The cost of cutting the cards for figured Wilton and of setting the loom for figured Axminster is not justified unless 500 square yards of carpet can be produced at one time, and unless 1000 square yards at the very least are sold. In Wilton ware, 1000 square yards may be taken as a reasonable sale, but in Axminster nothing under 10,000 or even 20,000 square yards is considered to be a good seller. In these circumstances, manufacturers shrink from abandoning anything that has sold before, and leave the trying out of new lines to hand-knotted goods where a failure hits one individual product only, and where overhead expenses are small.

However, this argument explains only why the industry cannot easily emerge from the abyss, not how it slipped into it. In looking for the actual reasons one may point first to the fact that the trade has been for a long time in the hands of the same few families, the expensive plant required hindering the foundation of new, more experimentally minded firms. One may also point to the prosperous condition of the carpet industry—no crisis, no sudden decrease in employment, no struggles for life have made it necessary to launch out on exceptional, outstanding enterprises. One may point to the atmosphere in which the works' designers have to live, shut up in their studios in a small and provincially minded town. Manufacturers feel this lack of brilliancy in their studio products, and for this reason buy sketches from outside artists as "food for the studio" (this was the expression used by a works' director). But here we are faced with another evil. For one thing, a public designer whom I met told me that he did not believe there were any colleagues of his (he did not exclude himself) with other than a selling interest in their designs. This is not difficult to explain. Even granted that there are some quite pleasant sketches bought by manufacturers, what happens to them when they have got into the grip of the studio? They are "mashed up", as the hearty term goes, and not merely redrawn on point paper. You might have a nice all-over pattern, say, some spots and waves in various shades of light grey, and at the end of this uncanny process you will find that the pattern has been used as a centre motive and again in the four corners, that the details are cut out, and that the greys have now become brown and the inevitable orange. A young manufacturer, showing this procedure to me, prophesied that at the end I should hardly be able to recognize the original sketch. He was right.

But why is this practice necessary? Can the commercial artist not produce a suitable design at once? The answer seems to be that some, above all the German designers, do, but with such a loss of freedom and inspiration as to make those finished designs less popular with manufacturers. On the other hand, why does the staff designer not respect the character of the sketch instead of depriving it of as much charm as it may possess? Here, I am afraid, one has to refer again to the uninspiring life in a factory studio which means losing touch with good modern design.

So far the attitude of the manufacturer himself has not been taken into account, though this is no doubt the real root of the trouble. I do not wish to make vague and general accusations, but I can quote what manufacturers have said to me in front of their products at their showrooms. One man spoke of the modern "bastard stuff" which he has to turn out; a second said that he, for himself, would only buy genuine Oriental rugs; and a third said in a forlorn way: "I wish you could tell me who the people are who buy my stuff. I have never been in a house with carpets like that." There was also a fourth director whose expressions in front of his products varied between "hideous", "horrible", "beastly" and "nasty", and a fifth who seemed to get a perverse joy and self-tormenting pride out of displaying his worst best-sellers. I still remember the sound of his: "Now look at this, isn't it a brute?"

In vain I tried to discuss the question of practicable improvements. No manufacturer denied that simple modern designs would actually be somewhat cheaper in production than the usual elaborate concoctions. What they all insisted upon was that the current designs are what the public likes, and that customers would not buy anything else. My answer was twofold. On the one hand, it is a fact that the trade in plain carpets is growing. In two rather expensive departmental stores in London the sales of plain goods come to 45–50 per cent. and over 50 per cent. This certainly shows that there is a distinct demand, at least by a certain class, for something less offensive than the patterned carpets at present on the market. On the other hand—and now an argument has to be repeated which I had to mention in more than one of the previous sections—how can the manufacturer know what the public wants? In carpet shops, there is a habit—as a rule quite unconscious—of putting those designs on top of the carpet pile which the shopkeeper thinks he may sell and wants to sell. Less hackneyed and sounder designs disappear to the bottom of the pile, and are thus not seen by the majority of the customers.

For the moment we cannot go further. These few indications of the personal attitude of manufacturers and retailers must suffice. There will be opportunities of reverting to the problem in later pages.[1]

[1] A few notes may be added on the artistic situation in the *linoleum* trade. They are based on information received from one manufacturer only, and on the usual display of linoleum in department stores. Again we find an almost complete absence of satisfactory patterns. This applies to linoleum—a mixture of pulverized cork and oxidized linseed oil pressed on a jute canvas cloth—as well as to cheaper floor covers such as floor-cloth which is just a jute ground covered with several coats of oil colour. Again we find that only a few firms are of any commercial importance, in fact only about five or six in this trade. Again their designs come from works' drawing offices or public designers. Nothing can be compared with the admirable enterprise of some German linoleum manufacturers before the war who asked outstanding modern architects such as Peter Behrens to design for them. Patterns in Britain are either modernistic or Oriental. They last for two years or more. Changes seem slower than in the carpet industry. In the firm which I interviewed only 10 per cent. of the running designs are as a rule this year's. The demand for plain linoleum is considerable. If mottled shades are included it amounts to about 50 per cent. or up to 70 per cent. of the total output. Completely plain colours are rarer (about 10 per cent.), because they are supposed to show footmarks more easily. It is unnecessary to enlarge upon conclusions from these few facts. They are identical with those arrived at in discussing carpets.

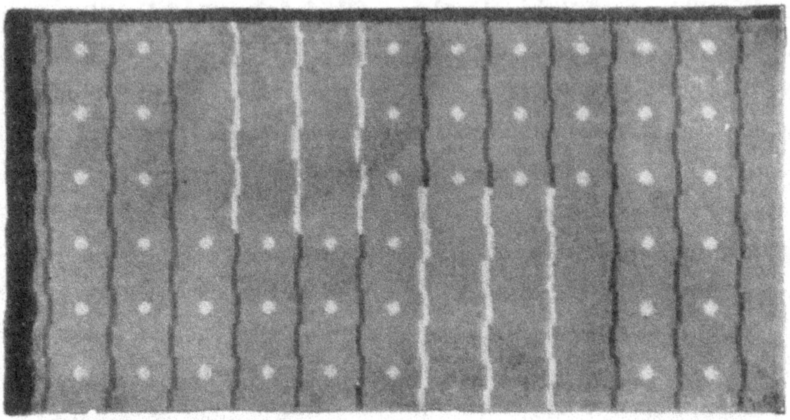

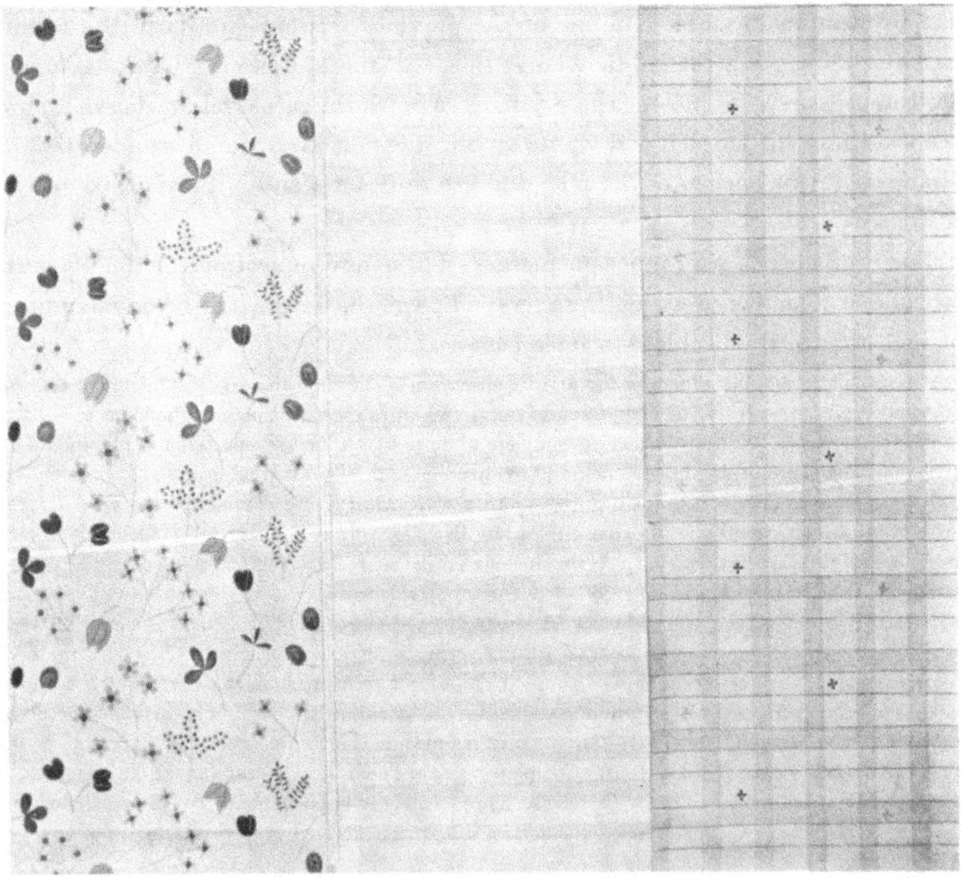

XII. *"Aquamarine", machine-tufted rug, made by Tomkinson's of Kidderminster to an adapted design of Marian Pepler.*
N.D.K. wallpapers (by courtesy of Gordon Russell Ltd.).

WALLPAPERS

There is one exceptional feature in the English wallpaper industry which may in our connection be of importance. About 90 per cent. of the trade is controlled by one combine, the English Wallpaper Manufacturers. I visited four firms: one was among the biggest units of the combine (about 1000 employees), the second did not belong to the combine but worked in close co-operation with it (about 800 employees), the third was a medium-size independent factory mainly producing medium and high-price machine-printed papers, the fourth specialized entirely in handblock-printed papers with period patterns. Here no new designs are required at all. The range of patterns and of blocks is regarded as closed. Some of the actual blocks in use date back to the first half of the eighteenth century. In addition, I saw an architect of the younger school who has, since 1933, designed some wallpapers which are carried out by a small manufacturer.

The conditions of design and the relation of design to production in the wallpaper trade resemble in many ways those which we have found in printed textiles. In both trades volume production is based on the process of roller printing, whereas high-price goods are handblock printed. In both trades, therefore, the initial expense is high for large-scale manufacture, and comparatively low for printing on a small scale. The engraving of rollers for a patterned wallpaper costs about £40, for an expensive embossed paper to be manufactured by machinery about £100. To arrive at the total preliminary expense, one has to add to the cost of rollers the artist's fee and the number of rolls used for samples. Thus a sum of £150 or £200 has to be risked on each new pattern before it begins to sell. No wonder that in these circumstances most manufacturers try to limit their range of machine-made papers to patterns which are sure to meet with a fairly large demand. An inexpensive or medium-price paper of which less than 5000 rolls are sold is considered a failure; 10,000 rolls a year may be regarded as a good, 20,000 as a magnificent result for medium-price ware. In cheap papers sales are expected to be still higher. Once more it is essential to realize the social meaning of such figures: 20,000 rolls—that means anything from 1800 to 2500 rooms which, in one year, are hung with that one particular paper. It will surround 1800 or more English families for five to ten years.

For new designs the wallpaper trade draws from the same supplies as the textile industry: staff designers with or without special art-school training, and commercial designers. A medium-size factory produces well over a hundred new designs a year. This figure does not include different colourings of the same design. In big firms the corresponding figure is sometimes much higher. Most of the designs are kept only for one or two years. In a large shop I was told that about 65 per cent. of the patterns on show were this year's range. It is very unusual to repeat a pattern each year for four or five years. As in other trades, the tempo of change seems to be quicker since the war than it was before.

The proportion between factory-produced and bought designs varies as much as we have found in printed fabrics; 50 : 50 may be the proportion in one firm, 35 : 65 in another, 65 : 35 in a third. Staff designers are paid at about the same rate as in the textile trade. £400 to £600 seems the average salary, but it may in a few exceptional cases go up to £1000 and more. The head of a wallpaper works' studio is as a rule fairly free in his production, although some (or perhaps most) of the general ideas underlying his designs may come from directors or sales managers. As to free-lance commercial designers a big factory may on an average have two or more calling each week. The majority are foreigners, that is representatives of big Continental studios. Germany supplies more designs than any other country. Even Old English chintz patterns for wallpapers are sometimes produced in Berlin. Fees are about 5 guineas for a sketch, and 8 to 10 or up to 15 for a finished drawing. The same studios design printed fabrics and wallpapers, and it is not rare that identical patterns are sold to both trades, in spite of the obvious fact that a wallpaper design should be conceived in terms of flat extension, a design for curtains in terms of creased draping.

There are hardly any independent artists or architects outside the trade who design wallpapers. The architect whom I saw is at present probably the only one in England. His papers are very pleasant in design, and certainly the best modern papers produced in this country. Not all of them are independent of Central-European influence, but some appear entirely original and distinctly English in their restrained patterns and colours. They are all handblock printed and consequently sold at about 5*s*. to 6*s*. a roll. Although their standard of design is evidently far higher than in most of the factory-produced ware, bigger manufacturers shrink from calling in architects or artists. They say that their outlook is too uncommercial and their taste too much at variance with popular demand. It is difficult to contradict such a statement, but one should not forget that about fifty years ago the leading English architects and designers—Morris, Voysey, and Crane for instance—invented wallpaper patterns and sold them successfully. If Morris and Co., Essex's and Jeffrey's could sell the designs of the then most advanced artists, why should the same not be possible now, if it were tried? Is the modern style so much less popular than the style of the Arts and Crafts? The artistic situation of the wallpaper industry in Germany tells against this supposition. The best range of wallpapers produced in Germany, the N.D.K. (Neue Deutsche Künstler) Tapeten, are designed by independent architects and artists like Professor Hillerbrand. The earliest and best of the so-called Bauhaus wallpapers were designed by artists who belonged to the Bauhaus. And both these ranges are produced on a large scale and sold at relatively low prices (about RM. 1.00 to 2.00 a roll).

The prevailing conditions of design in English wallpapers are closely linked up with their artistic value. This has surprisingly changed during the years of my research. Modernism began to burst in about six years ago. Floral period designs were given up and replaced by riotous pseudo-cubistic patterns derived from German "Expressionism", but devoid of the expressive qualities which were this artistic movement's *raison d'être*. After a few years colours were softened down, and jagged patterns in pastel shades became the fashion. The next step was that even these appeared too obtrusive. Manufacturers and the public realized at last that a wallpaper should be a background and nothing more, and designs were introduced the appeal of which depended exclusively on broken, stippled effects. The first of the Bauhaus papers were of that kind. In England they appeared about 1932 or 1933. Naturally, they give much less scope to the individual designer than floral or cubistic patterns. Most of the broken effects are the result of experiments with rollers, and the result can never be exactly foretold. However, it seems as though at this promising moment the disappointed wallpaper designer must have looked for a new outlet for his ornamental fancy, and found it in what is called "appliqué" work. There had been a vogue during the last years before the war to imitate the appearance of eighteenth-century panelled walls by applied borders with period motives around plain wallpaper panels. This scheme was now modernized and most enterprisingly developed. Instead of the simple and harmless borders of pre-war days, one was offered clusters of stylized leaves in brown, green and orange (the notorious colours of the modernistic carpet), or clusters of odd stalactitic shapes to be stuck along the upper edge of the panellings and especially in the corners. Sometimes appliqué grew bolder than this, and life-size hollyhocks or gondolas swimming in a printed and cut-out sea in front of printed and cut-out palaces rose in profusion from the bottom of the panelling. When I began my research this was still going strong and seemed to endanger all the gratifying results which one might have expected from the new "textured" papers.

However, the appliqué craze is passing now. In the cheap trade it still lingers on, although in less wild forms. The builders of suburban semi-detached houses still feel it their duty to provide at least the "best rooms" with some appliqué work. In a big firm I heard that about one-third of all the papers sold to builders or the public is used with some sort of patterned border. In the medium and higher-price trade plain textured patterns began to outstrip cubistic patterns about two years ago, and are now indisputably the best-selling type. The demand for big "Modern Floral" all-over patterns (as a rule something like brown and orange autumn leaves) has become very small, nor is there much demand for period floral patterns of the chintz character, although this is supposed to have shown an upward tendency within the last six or twelve months. By no means all the textured designs are good. Many of them are rather crude and thoughtless, and very few have the delicacy of the first Bauhaus designs. However, looking back at what they have replaced, one has duly to appreciate that they represent a considerable improvement.

Yet it remains once more a deplorable *testimonium paupertatis* that such an improvement could be attained only by ruling out real pattern effects altogether, instead of by the introduction of better patterns. Except for the few designs by the architect whom I mentioned very few wallpapers exist in England that can be compared with the charming and delicate designs of some of the expensive Swiss Salubra and most of the less expensive German N.D.K. wallpapers. Is this due to an inborn lack of appreciation in the English public, as most of the manufacturers want to make you believe? Certainly not, for there have been the Morris and the Voysey papers, and there are all the pleasant patterns on printed textiles. Thus the only reason to be adduced seems to be the absence of creative personalities in the studios. The employment of modern artists or modern architects would no doubt be a great help, as it has been to the textile trade by the work of such firms as Allan Walton. It is a question worth some thought, why the wallpaper trade has not availed itself of the same artistic forces which have so obviously been to the benefit of the textile industry. I received one remarkable answer to this question from two separate informants: Crane, Voysey and other outstanding artists who designed wallpapers before 1900 worked for a few medium-size firms, the owners of which were personally interested in good design. When the combine was founded in 1899, this personal note of necessity disappeared. Also it was soon realized that there were cheaper methods of obtaining designs than by buying them at 12 or 20 guineas from famous artists. The changed policy did not have an unfavourable effect on sales, and as the policy of the combine meant the policy of almost the whole of the trade, architects and artists soon found it difficult or impossible to sell wallpaper designs to manufacturers. I repeat this story as it was given to me. So far as I know the individual units within the combine are fairly free in their selection of designs. Besides, there are still some independent firms. Thus one can still hope for an Allan Walton or an Edinburgh Weavers of the wallpaper trade who may come and take advantage of the imagination of existing architects and artists for a revival of good contemporary wallpaper pattern.

POTTERY

A considerable portion of the English pottery trade is located in the Five Towns. Stoke means more to the potting industry than Kidderminster to the carpet industry or High Wycombe to the furniture trade. The largest works weaving carpets are outside Kidderminster, the largest cabinet-making factories outside Wycombe. But the Five Towns house the biggest as well as the smallest firms, about 400 altogether, giving work to some 60,000 or 70,000 employees. I visited fifteen firms, employing about 10,000 workers. One of the factories produced lithographs for transfer on to pottery—a process to be discussed later—one specialized in insulators for telephone, telegraph, electric railways, etc., one made, apart from tableware and kitchenware, sanitary appliances on a large scale. Two of the factories seen were of a definite mass-production character, one of them employing more than 3500 people, the other about 1500. One manufacturer (outside Stoke) produced expensive porcelain only (300–400 employees), the rest either china and earthenware or earthenware exclusively.

It is well known that earthenware plays a more important part in English than in Central European pottery. Cheap porcelain scarcely exists in this country. The difference between china and earthenware lies, as is well known too, in the clay as well as in the firing. English earthenware roughly speaking contains 20 per cent. China clay and 15 per cent. Cornish stone. The rest is ball clay and calcined flint. The usual firing temperature is 1000–1100°. English bone china is fired at 1200–1300°, and consists of about 25 per cent. China clay, 25 per cent. Cornish stone and 50 per cent. calcined bones. Continental china of the Dresden type requires a firing heat of 1350–1400°, and is made of 50–65 per cent. China clay and 30–35 per cent. felspar, the rest being quartz. England's famous potters in the eighteenth century, such as Wedgwood and Spode, produced earthenware only. And still to-day 80 per cent. of Wedgwood's and 70 per cent of Copeland-Spode's output is earthenware.

Among the earthenware and china works which I visited, and which have some artistic ambitions, the largest had a staff of 600–800, the smallest less than 100 employees. One of them was the studio of a potter whose standards of craftsmanship are of the highest. In another the work was confined to excellent modern decoration of bodies bought from mass-production factories, a system exceptional in the potteries but reminiscent of the division between producer of cloth and printer in the cotton industry. The conclusions which I deduced from my interviews with the directors of the fifteen firms have been checked by information from several shops and stores.

The statements of retailers were especially useful with regard to the order of popularity as between the various types of pottery decoration in the English market. Manufacturers and distributors classify these as a rule under All-over Prints, Period Floral, Modern Floral and Banded. The term "All-over Prints"

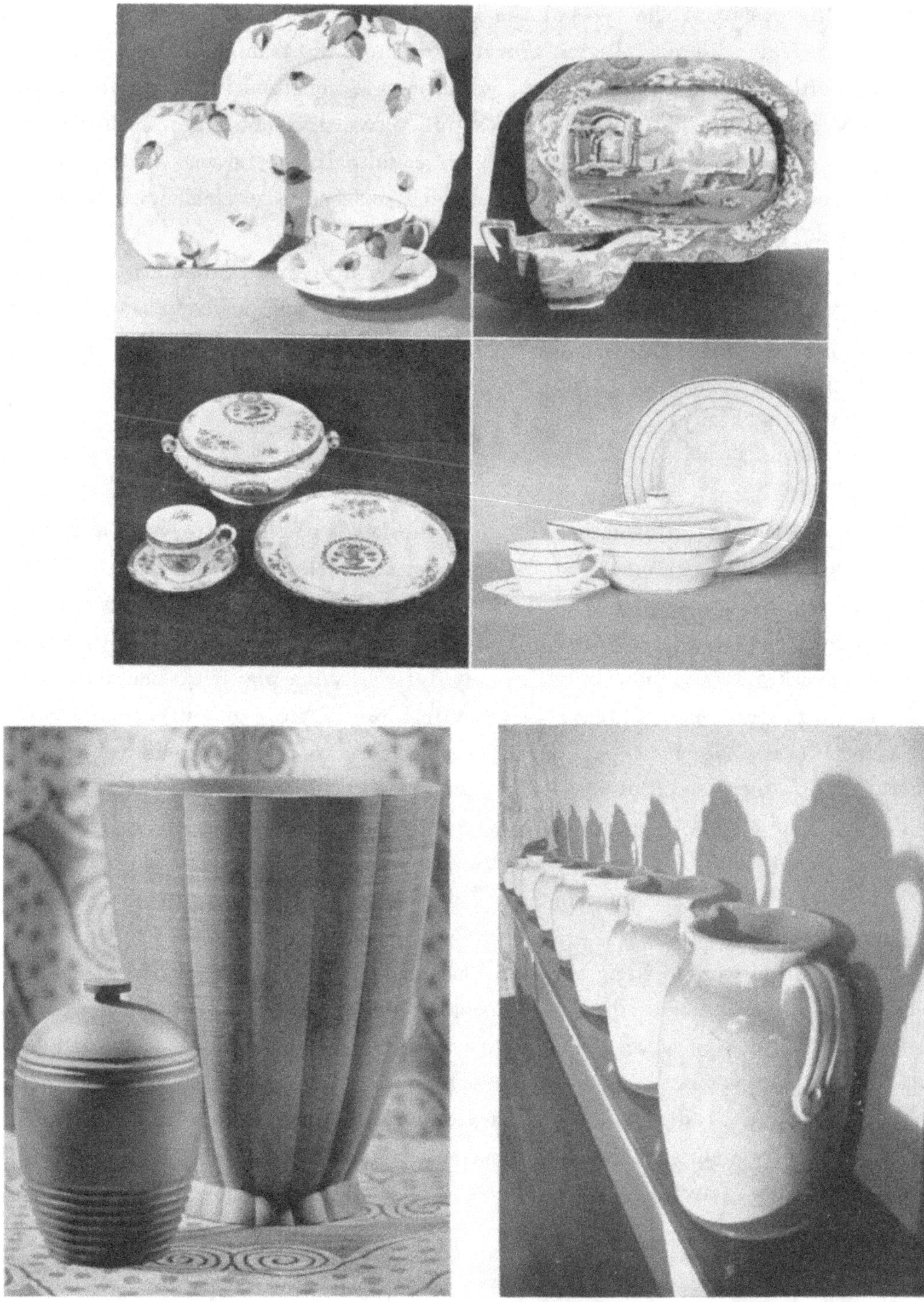

XIII. *The four most usual types of pottery decoration: Modern Floral; All Over; Period Floral; Banded. Vase and tobacco jar designed by Keith Murray and made in bronze basalt ware by Wedgwood's. Acid jars made by Royal Doulton.*

means any design of the type of the "Willow Pattern" or "Blue Italian" or "College Views", etc. "Period Floral" are those old-fashioned floral borders, spray borders, broken borders, etc., which are to be found in most eighteenth-century porcelain. "Modern Floral" signifies two different types, either hand-painted rustic designs or lithographed (or hand-painted) flowers of a modern commercial character, often intermixed with some cubistic elements. The term "Banded" needs no explanation.

The managing director of a big china shop with several branches in the Midlands gave me the following index figures as to their sales early in 1934:

All-over Prints:	Dinnerware	4	Teaware	7	Ornamental	3
Imitation Period (prints chiefly)	,,	11	,,	9	,,	4
Modern Floral	,,	37	,,	41	,,	48
Banded	,,	31	,,	14	,,	11

Provided with this information, I visited the heads of the pottery departments in three London departmental stores, in order to check the figures quoted above. In one, which has proved remarkably go-ahead in recent years, I was told in 1934 that they sold hardly any of the commercial Modern Floral, because their buyer dislikes it too much, but that they did well with plain (40 per cent.) and banded (25 per cent.) pottery. The Rustic Floral came to about 10 per cent., Period Floral to about 20 per cent., All-over Prints to 5 per cent. In 1936 plain had slightly increased (now 45–50 per cent.) above all in the higher price ranges. Banded had slightly decreased (now 20–25 per cent.). Period designs have increased in more expensive, decreased in cheaper ware (now 20 per cent. altogether). All-over Prints come to 5 per cent., Rustic Floral to hardly more than 5 per cent.

In the other store in 1934: All-over Prints 10 per cent. or less, Period Floral less than 10 per cent., both selling to old people only, and Banded about 25 per cent., Modern Floral even more. Plain pottery was not mentioned. In 1936: Banded about 50 per cent. in dinnerware, about 25 per cent. in teaware. Modern Floral more than 50 per cent. in teaware. Rustic Floral has lost ground. In dinnerware, Banded now amounts to about 50 per cent., whereas in teaware, Modern Floral comes to more than half the sales, and Banded to about a quarter. Among the

Modern Floral, the rustic type has decreased, and is now considered a little "old-fashioned".

In the third store, the most expensive of English and probably of European department stores, the situation was remarkably different. Their clientele, the bulk of which earns more than £1000 a year, wants period porcelain and earthenware more than anything else. This comes to at least two-thirds of their sales, Banded to perhaps 15 or 20 per cent.; Modern Floral is without any importance. In the best-known special shop in the West End, Period Floral is also the dominant feature. Only very little Banded and Rustic Floral is on show.

This should be compared with the West End branches of the most popular multiple store. Here the display consisted in 1936 of much plain white, several vulgar floral patterns, a few rustic hand-painted sets, one of which was of quite a pleasant design, one All-over Print, and some Banded sets, either with traditional borders or with simple modern bands.

The relation between class of shop and type of design in vogue which is evident in the last cases mentioned, deserves some closer consideration. It runs parallel to another relation, that of potting material and type of design. The best and the worst decoration we find in earthenware. Above a certain level of price, Modern Floral decreases rapidly in importance. On the other hand, Banded patterns are typical of medium-price ware. They go hardly up to the most expensive or down to the cheapest market. A manufacturer of picnic cases confirmed this. Though his Banded plates, cups and saucers were actually cheaper than his ugly Modern Floral ones, he had to put the Banded ware into the more expensive cases, and the Floral ware into the cheap ones in order to satisfy his various markets. However, good or bad, Banded or Floral, English earthenware is by now modern (or modernistic) in appearance. It was about eight or ten years ago that commercial Modern Floral forced its way into the British market. Banded patterns came a little later, about 1930, and remained confined to only a few firms of artistic ambition until 1933. Since then it has, as was shown, steadily gained ground.

It is different with porcelain. Here the trend has changed very little during the last twenty years. A well-known firm of eighteenth-century fame gave me as their ratio of production: almost 50 per cent. All-over Prints, 30 per cent. Period Floral, less than 20 per cent. Banded. A manufacturer of cheaper porcelain said also that the majority of his output was period ware, and only 15 or 20 per cent. Banded; and in the showrooms of the two most exclusive pottery works in England you see hardly anything but period imitation. I asked the managing director of one of these two why this was so and received a remarkable answer. He said that in England china is bought only by the wealthier classes. Now the young members of these classes prefer spending their money in motor-cars, extensive cruising, and other more up-to-date pleasures. The new habit of outdoor week-ends, as well as the size of the new West End flats, also have a share in reducing the market for expensive porcelain. Thus the customers to cater for are all older people of a decidedly conventional taste. This is quite different in those Continental countries where porcelain is less costly and more usual. Here there is more scope for progress in design of china, and even manufacturers of such established character as Meissen or Berlin have introduced some modern lines. The plain Berlin set "Neu-Berlin" is a remarkable example of German high-class china, modern without any decoration, beautiful in shape, original and yet not strikingly novel.

So far as I am aware no such experiments have yet been carried out in England, although it is by no means proved that there would really be no market for unadorned modern china. A change of taste in Court circles could, for instance, create a considerable demand almost at once.

But do we really know about demand? Again we are faced with the problem that went through the discussion of all the previous sections of this book. Again the manufacturer argues that he is following the public, and the public argues that it can only buy what it is offered. Again the distributor is in between, acting as interpreter between consumer and producer. How little has been done to collect reliable information on popular taste. My questionnaire at the Midland Industrial Art Exhibition did not prove much; but what it seemed to prove was the responsibility of the manufacturer. One group has for years now promoted good and simple design, another with equal fervour meretricious modernistic decoration. For period forms, there has been hardly any propaganda. Consequently of fifty-seven answers received, four declared an almost irreproachable banded pattern their favourite, four another more conventional banded pattern in green and silver, but four also a hideous modern floral set of no character whatsoever. Twelve votes, on the other hand, were unanimously against one floral monster, nine against an insignificant period copy. This result may be safely interpreted so as to attest the educational success of the one group, but also the partial invalidation of its efforts by the other group.

Another noteworthy case was an exhibition held at Harrod's in 1934. Its origin deserves to be recorded. Mr T. A. Fennemore, then a director of Foley China, had asked twenty-seven artists to design new patterns for him. They were paid £10 a design plus royalties, and the designs were shown at the exhibition and later put on the ordinary market. Among the artists invited were representatives of many opposed tendencies, as is proved by names such as Frank Brangwyn, Laura Knight, Albert Rutherston, Vanessa Bell, Duncan Grant, Paul Nash, Ben Nicholson. Some of them sent only one, some two or three designs.[1] Unfortunately, the experiment was limited to new decoration and did not include new shapes—which was obviously a grave drawback. But as the manufacturer pointed out, he could not have afforded such an experiment at all, had it entailed the expense of new moulds for more than thirty sets.

I shall have to refer more than once in later pages to this interesting experiment. In our present context, its importance is that the most popular sets were almost all gaily decorated with figures, fruit, etc. Now it must of course be admitted that here again the validity of the results was limited for various reasons. For one thing, prices were so high as to exclude the majority of the buying public. They ranged from £1. 15*s.* to £5 for a tea-set of twenty-one pieces. Of thirty-three designs, twenty-eight were over £2. 5*s.* It must also be borne in mind that in purchasing such "first editions" (as they were called) there may have been a certain amount of snobbery. On the other hand, new tendencies very often start at the top and only slowly filter down. So it may be that the Foley Exhibition can be taken as the symptom of a growing demand for more elaborate and at the same time distinctly modern decoration.

Supposing that we can take this for a moment as granted, what would be the consequence in the ordinary market? How would it work out? It must not be

[1] For illustrations, cf. H. G. Dowling, *l.c.* pl. 37ff.; also *Commercial Art and Industry*, vol. xvii, pp. 223ff., 1934.

forgotten that the Foley sets were designed by independent and distinguished artists. But who would design them as a rule? The use of free-lance artists is very rare in the pottery trade, although it has proved a system yielding remarkably good results. Besides the Foley experiment, one has to point above all to Wedgwood's who have employed a number of artists in some way or other for quite a long time. Their most successful alliance is that with a well-known architect and designer in London who spends two months' work on pottery design. Apart from a fair salary, a royalty is paid to him for all sales over a certain limit which is fixed considerably higher for tableware than for vases, bowls, etc. Every piece made from his designs bears his name as well as the manufacturer's. This arrangement seems to work very well indeed. Both firm and artist pronounced themselves satisfied; and the style thus introduced has proved its vitality by calling forth many imitations from various other factories.

These imitations are due to staff designers or so-called decoration managers. The employment of full-time designers is less a matter of course in the potting industry than the public might expect. Several times it was stated to me as an established fact that 90 per cent. of the firms in the Five Towns have no proper designer at all. The decoration manager is usually a man brought up in the trade, in the appalling atmosphere of that shapeless agglomeration which is now called Stoke-on-Trent. He may have attended evening classes in an art school, or he may not. Sales experience is regarded as more important than a grasp of design. So far as design is concerned, his work consists of the buying of lithographs for transfer, the adapting of hackneyed old patterns, and the imitation of new ones from next door.

Payment is in accordance with this state of affairs. The decoration manager in a small works may get about £150 a year, or anything up to £250 or £300. Firms of some commercial standing pay an average of £300–£500. There are hardly more than six or eight manufacturers in the trade who value their head designer so highly as to give him a salary of £600 or £700. One man was said to get £1000. I have not met him.

When I asked manufacturers to give me a reason for the low salaries of artists in an industry for which artistic design is evidently of more importance than in most other trades, the answer was usually this: Though the ceramics industry requires an especially large number of designs—to give just one instance, the lithographic transfer works which I saw produces 400–500 patterns a year—the volume of trade is very small, if compared, say, with textiles. If we except the three or four biggest factories—of which one gave me as their output 750,000 pieces a year—we get down to figures like 1000, 2000, 5000 sets a year for good sellers. So designers are kept very busy; but the returns from their individual designs are not anything like those from designs for Manchester cotton fabrics. This being so, gifted and enterprising students of design tend to try for jobs outside this trade.

Still many of the good designers in the ceramics industry appear to have gone through the Burslem School of Art which, under Mr Gordon Forsyth, has gained a considerable influence over a certain section of the trade. Mr Forsyth, who has worked in factories himself, is especially interested in the hand-painting of pottery, and has, in spite of the inadequate equipment of his school, succeeded in training a great number of students to carry out deftly, or even to create, floral patterns of a pleasant rustic type. Thus he has given the ceramics industry a new impulse towards a sound technique of pottery decoration. This technique, as opposed to the hand-colouring of engravings which is usual in the production of period ware (up to 75 per cent. in the porcelain works that I visited), means a reintroduction of creative craftsmanship which corresponds to the contemporary revival of artistic craftsmanship in other trades, the revival which was originated by William Morris and his disciples.

However, hand-painting and engraving are not the only methods of decoration in use in the ceramics industry. The most widely adopted technique nowadays is the transfer of lithographs on to pottery. This was introduced into England shortly after 1880. Of those whom I saw, the biggest mass producer told me that, of his output, 40 per cent. is plain, 10 per cent. white and gold, 35 per cent. decorated with lithographs, and 15 per cent. engraved. Hand-painting is scarcely done. This is a characteristic fact. We find that the largest firms prefer lithographs, the smaller firms hand-work. The economic reason is obvious. Hand-painting requires no outlay of capital, though the wages of the hand-workers (30*s*. to 50*s*. a week) are slightly higher than those paid to workers who are responsible for the transfer of lithos. Lithographs are produced in special works and sold by the sheet.[1] The complete edition of one pattern consists of 2000 to 3000 sheets, and unless a pottery manufacturer buys the whole edition up, he is not safe from competitors. Therefore it pays only to use lithographs if about 50,000 to 60,000 pieces of one design are going to be produced. This means, of course, considerable initial expense.

The designing is, as a rule, done in the printing firm. Only one out of five patterns, I heard, comes from the pottery studio. There is no technical reason for this, and it seems to me that the aesthetic results might be greatly improved by increasing the potter's share in the invention of the lithographed patterns. I know of only very few cases where adventurous potteries have begun to design modern lithographs to satisfy their standards. Miss Susie Cooper, so far as I know, was the first to do this. Apart from hers and a couple of others (Pountney's of Bristol), I have not seen any acceptable lithographs, and one of the foremost experts on English pottery, a member of all the committees, confirmed two years ago that "in the whole of the Five Towns, there is not a single decent lithograph." In fact, the director of the lithographing firm whom I saw admitted unhesitatingly that he regarded many of his patterns as hideous.

This state of affairs constitutes almost the worst obstacle to any thorough artistic improvements in the ceramics industry. For in the cheaper trade, lithographs are dominant. In one of the department stores which I saw, about 50 per cent. of the pottery in stock was lithographed. It is all the more essential to draw attention to this matter, as it is almost completely unknown to the public.

So far we have discussed only pottery decoration. But in the ceramics industry the design of shapes certainly plays an equally important part, or at least it should do. In point of fact, manufacturers are far readier to turn out new patterns than new shapes. The reason for that has already been mentioned in connection with the Foley Exhibition. It is the changes of shape as a rule that are expensive and not those of decoration. Some more technical remarks may serve to make the meaning of this clearer.[2]

To-day only a few factories and studio potters throw on the wheel. In most

[1] There are four big litho-printers in the Five Towns. I visited one of them.

[2] For the technique of potting, cf. Gordon Forsyth's excellent and abundantly illustrated book, *The Art and Craft of the Potter*, Chapman & Hall, 1934.

works objects are either finished on a lathe, or jolleyed, or cast—to mention only the most important methods. Firms of high standing throw vases, bowls, etc., in the rough, and then stick the moulds to the lathe. All the final shaping is done with a steel blade handled by a skilled craftsman. Cups, saucers, plates, etc., are usually jolleyed. In this case, the clay is flung into a mould which is fixed on a revolving shaft. The shape of the mould corresponds to the exterior form of the object to be made. Then the jigger, i.e. a kind of blade carrying the profile of the interior, comes down, revolves and finishes the process. This is, of course, an operation "entirely devoid of any skill on the part of the operator" (Gordon Forsyth). The same applies to casting, which is used for pots, etc., and for most cheap commercial production. These technical considerations largely account for the fact that changes of shape occur less frequently than changes of pattern. Whereas the average pattern—except for some old-established ones—lasts two or three years only, shapes may be repeated for decades and more. There is nothing basically wrong in that. The cup, the plate, the coffee-pot are objects so unchangeable in use that only small variations of shape can be introduced without reducing their practical value. Thus Wedgwood's are still producing their eighteenth-century shapes, and some of them are so perfect that it would be unjustifiable to abandon them. One of the best contemporary sets, W. Moorcroft's famous Plain Blue, was designed in 1913 and is, in spite of that, as "modern" as anything created now, and as "modern" as Josiah Wedgwood's sets, i.e. undatedly perfect. The same thing may be said of some English eighteenth-century cutlery or lettering.

It is understandable that in such circumstances the best English manufacturers have been anxious not to depart from their sound traditions. Not that new shapes are rare on the market, but they are almost all thoughtless in design, and hardly meant to be of lasting value. The British Industries Fair, *The Pottery Gazette*, and big china shops display any amount of jagged handles, square or polygonal plates, wrongly balanced pots, etc. The main point (in our context) is that amongst English ceramics of artistic standard new departures in shape are rare. This is undoubtedly due to the national character of the English. In Germany and other countries, which are more *novarum rerum cupidae* than England, the perfection of the eighteenth-century shape would not have been a

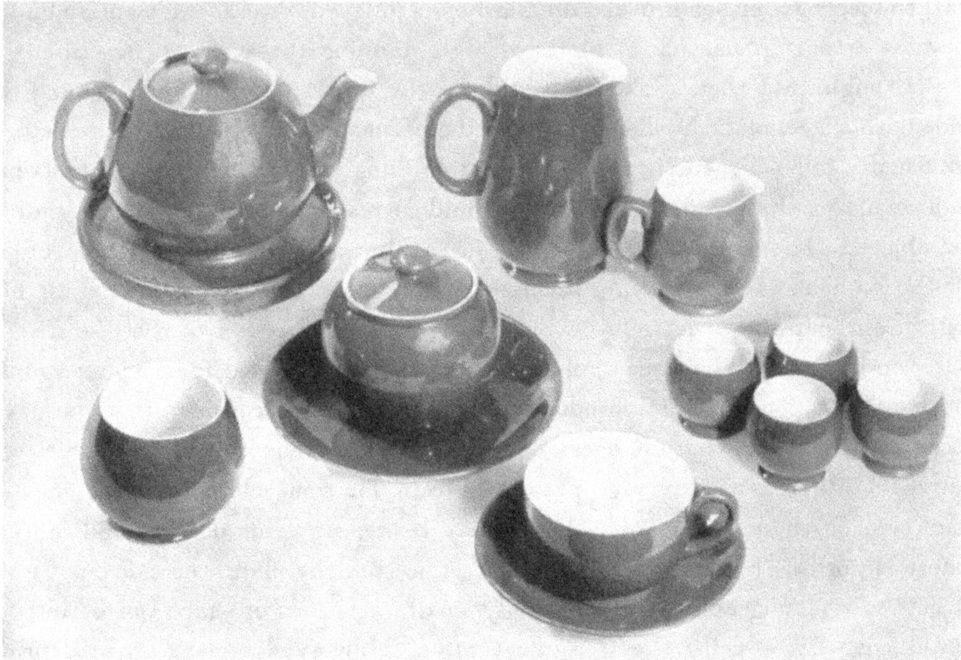

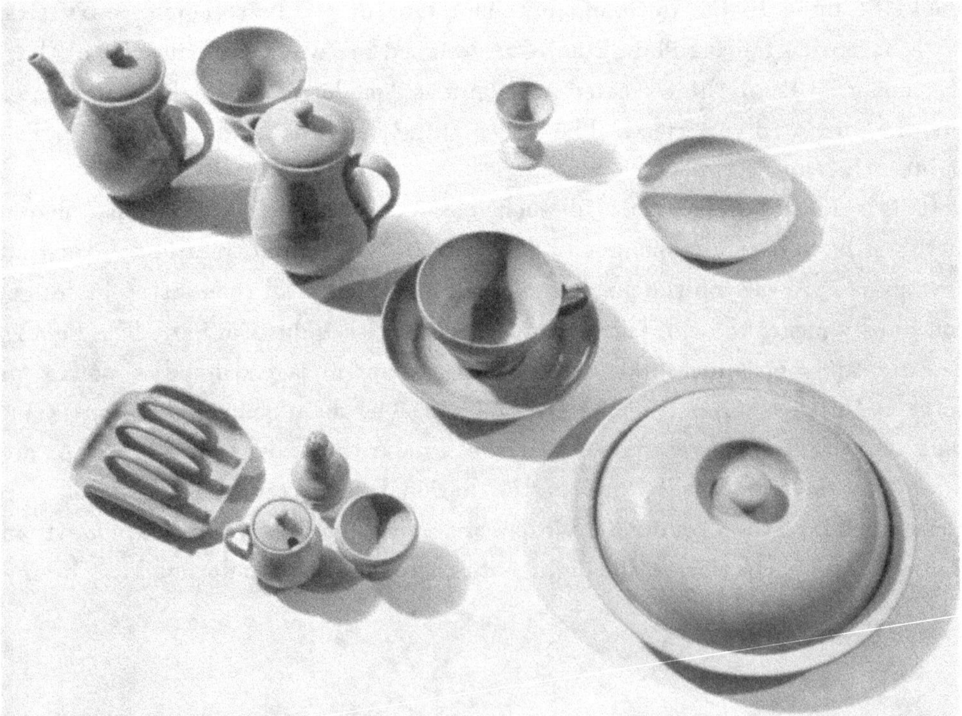

XIV. *Dark blue tea set designed and made by William Moorcroft.*
Some of the eighteenth-century shapes still used by Wedgwood's in conjunction with their modern shapes, and looking equally up-to-date and perfect.

sufficient reason for desisting from new experiments. The Bauhaus pots and cups may be less perfect than some of Josiah Wedgwood's, but they express one quality which Wedgwood of necessity could not bestow upon his objects—the spirit of the twentieth century.

Moreover the new processes of pottery mass production should raise new problems of design. For whereas to-day many of the factories still keep the same blackened sheds and the same kilns which they have had for more than a hundred years—another fact, incidentally, that shows how alive tradition is in the English pottery trade—there cannot be any doubt that the methods and requirements of a changed era are bound to react on the artistic appearance of pottery in the near future. Where casting with steel dies has replaced turning, certain basic qualities of appearance common to turned and thrown objects lose their justification. The initiative passes over from the artistic designer to the engineer. Little has been done so far to investigate seriously the influence of mass-production methods on the artistic side of ceramic products.

Yet pottery manufacturers know that mass production can in ceramics, at least as regards the quality of the body, help greatly toward improvements. The reason is that more efficient kilns can be used and better conditions of firing attained. Thus I was told on various occasions that even the best manufacturers can hardly compete with the strength and clearness of certain mass-produced and electrically fired wares. I know of only one case, where a particularly adventurous young potter has gone into partnership with a large mass producer, in order to take advantage of his non-intermittent kilns and of the best places in his ordinary kilns, which are not important in producing kitchenware and cheap table sets.

Another characteristic instance of mass production having led to highly satisfactory artistic results is the story of Royal Doulton's famous Acid Jars. When these were brought out, nobody thought of art. Soon, however, they were discovered by an imaginative architect and launched as objects for the adornment of the house. At present they are amongst the best-sellers in most shops dealing in well-designed modern goods. Somewhat similar is the case of some kitchen crockery, such as the blue and white Cornish kitchenware, or some sanitary pottery, lavatory basins and the like. As long as considerations of usability dictate the shapes, one need not be afraid of horrors. As with electric fittings and fires, it is the conscious application of art which tends to spoil the design. This is especially obvious in some of the most recent models of lavatory basins, etc., where simple rounded forms are replaced by cubistic follies. At Woolworth's too, the stoneware pans, etc., are the most satisfactory, the "fancy vases" the most hideous, of the objects displayed in the ceramics departments.

GLASS

The artistic situation in the English glass trade can be described in the following way: modern shapes and modern decoration are still exceptional. Until 1934 only two firms had done serious and persevering work in introducing good modern design; one of them a small works with a well-established tradition of craftsmanship, the other a larger factory enterprising enough to appoint on a part-time basis the same architect whom I mentioned as being one of England's best pottery designers. As for the rest, almost all that is produced by English factories is either old-fashioned richly cut crystal ("death by a thousand cuts") or cheap-looking pressed glass of no distinctive character. In 1934 and 1935 a few further experiments were started by mass-production as well as by high-class firms. The initiative was due to exhibitions such as that at Harrod's which I discussed in connection with Foley china, and those organized by the Royal Academy early in 1935 and by the Royal Institute of British Architects early in 1936. At Harrod's one manufacturer of crystal showed vases and tableware designed by several well-known artists; but as he refused to give me any evidence I am unable to say whether this departure has had any effect on his ordinary production. On the whole, it seems as if the only influence of the Modern Movement on manufacturers like the one mentioned just now has been a certain reduction of cut, and a certain preference for simpler patterns. Not that this development has gone far as yet. It remains a puzzling question, why the British glass industry has proved so stagnant and narrow-minded.

It is an old industry with as much tradition as pottery. But while tradition has proved beneficial in the ceramics industry, it does not appear as a blessing in the glass trade, at least from our point of view. The reason is that, in traditional pottery, the prevailing taste to-day is 1770, so to speak, whereas in traditional

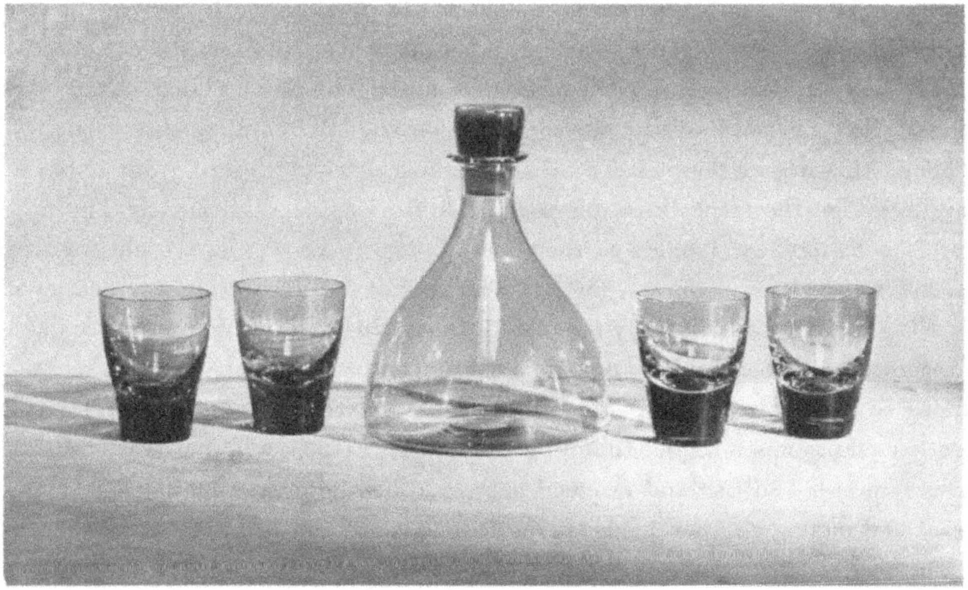

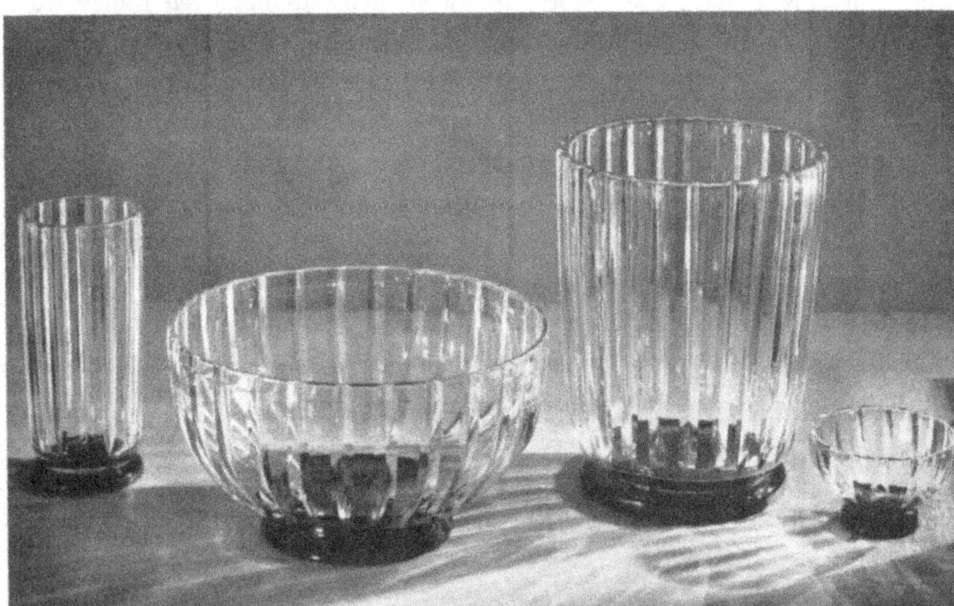

XV. *Sherry set. (James Powell & Sons Ltd., Whitefriars.)*
Vases and bowls, designed by Keith Murray and made by Stevens & Williams.

glass it is 1870. Another reason is to be found in the different history of the two industries. In pottery, the rapidly growing demand of the last sixty years was met by an ever increasing number of English firms. The production of cheap ware was started and successfully developed beside that of more expensive quality. Though this has no doubt led to a considerable deterioration of taste, it has at any rate kept the trade from stagnation. In the English glass industry, nothing corresponds to cheap banded earthenware. Inexpensive articles which you now see in English shops are mostly foreign. The bulk of the cheap trade was destroyed by the great depression of 1875–85. Since then England has no longer been able to compete with Belgium and Bohemia. Manufacturers did not want or could not afford to modernize their plants. Also the standard of living of the English worker was so much higher than in Bohemia that prices could not be kept on the level required. So England resigned herself to having only a high-grade crystal trade and buying cheaper goods abroad. In 1921, 6600 operatives were working in the Midland glass industry, as compared with 4500 in 1861. The corresponding figures in the brass and copper trade are 40,100 : 10,700.[1]

As things are to-day, the artistic standard of cheap imported tableware is remarkably high. A test of that is the glass department of Woolworth's. It contains hardly any article which is not at least acceptable, and many a design of considerable grace and delicacy. One cannot too strongly emphasize the fact that here the public is able to buy good modern design at 3*d.* and 6*d.*—and that it does so. It is one of the clearest denials of what so many manufacturers tell you about the hopelessly bad taste of the public. Of course this gratifying attitude of the consumer is partly due to the cheap prices and the tempting atmosphere of the Woolworth store. Still, it remains a fact to be borne in mind that customers do not seem to have left the stores in protest, in order to get something more traditional or more modernistic elsewhere.

However, English manufacturers deny that they could produce similar goods at a competitive price. Here and there it has been tried and was, so I gather, a commercial failure, because it is not easy now to make up for what has been neglected for over half a century. Factories in Central Europe are working with

[1] *Vide* G. C. Allen, *The Industrial Development of Birmingham and the Black Country, 1860–1927*, Allen & Unwin, 1929.

a much more modern equipment for the mass production of tableware and have developed a system of specialization that goes beyond anything attempted in this country. They have, moreover, at least in Czecho-Slovakia, the advantage of being able to give out a good deal of the blowing and decorating to home-workers, who are helped in this work by their wives and children. And since in the glass trade material and working process are comparatively cheap, while wages amount to 50 per cent. of the total production cost, the effect of the Bohemian system on the final retail price is obvious. To compete with this in England would mean such a large capital expenditure for the setting up of new plants that manufacturers do not want to take the risk.

While such considerations account for the absence of cheap British tableware, well-designed objects of a higher price have to struggle against Swedish competition. The recent development of the Swedish glass industry is one of the most remarkable chapters in the history of modern industrial art. Sweden is not by any means a country with an especially strong tradition of exporting glass. It was the independent action of a few firms which started the Swedish production of perfectly designed modern glass at reasonable prices. The "metal" used in Sweden consists of sand, carbonate of soda and carbonate of lime; the composition used in Venice is similar; in Bohemian crystal glass, potash is used instead of soda; in English crystal the principal ingredients are sand, red lead and carbonate of potash. The technique used as a rule in Sweden is that of blowing into wooden moulds, which is a fairly cheap process. In England this is usually followed by the cutting, whereas most Swedish tableware is left without any cut or engraved decoration, relying entirely on the beauty of shape. This is what makes Swedish glass so cheap. Venetian pieces are also usually unadorned. But their price is bound to be higher, because no moulds are used.[1]

The introduction of the new style in Sweden goes back to Consul Johan Ekman who, in 1915, started the production of "art-glass" at the small ink-bottle factory of Orrefors, by calling in two Swedish artists. In doing so he followed the suggestion of the Svenska Slöjdföreningen (Swedish Crafts Society), the organization corresponding to the German Werkbund and the English Design and Industries Association. The two artists were Simon Gate (appointed

[1] Some illustrations of glassmaking in *Commercial Art and Industry*, vol. xv, pp. 53 ff., 1932.

in 1916), a portrait painter and illustrator, and Edward Hald (appointed in 1917), also a painter. They were paid a certain percentage of the total sales of "art-glass", a minimum salary being fixed. In 1933 Hald was made a director of the works. Neither he nor Gate had done any glass designing before. In spite of that, their careful study of material and processes together with their high talents enabled them to create such convincing shapes and such enchanting decoration that within a few years the principal European markets were conquered.

The sales figures of "art-glass" produced at the two works of Orrefors and Sandvik are as follows:[1] "Art-glass" constituted 63·8 per cent. of the Orrefors and 55·2 per cent. of the Sandvik output in 1934, the rest being bottles, old-fashioned cut glass, etc. Taking the sales of "art-glass" for Orrefors as 100 in 1917 and for Sandvik as 100 in 1922, the figures for the following years are:

	O.	S.		O.	S.		O.	S.
1918	300		1925	1440	?	1932	1126	359
1919	264		1926	1710	?	1933	2078	454
1920	337		1927	1783	?	1934	2792	547
1921	135		1928	2470	654	1935	3285	589
1922	532	100	1929	2115	602	1936 (Jan.-Nov.)	3704	651
1923	1024	?	1930	2245	585			
1924	1151	?	1931	1655	487			

These figures should once and for all drive home the fact that a manufacturer's untiring and energetic activity for the perfection of design can lead to an immense increase in sales, provided he is willing to take the risk of doing something never attempted before, and doing it in an uncompromising way, and provided also that he is able to get hold of the right artists.

After a short time other Swedish firms followed Orrefors and shared in their success. It seems that Swedish glass was first introduced into England by Sir Ambrose Heal, who had been captivated by the display at the Göteborg Exhibition in 1923. Through the work of Heal's and an increasing number of other shops, Swedish glass became popular in this country and played a considerable part in making the public aware of the beauty of modern design in goods of everyday

[1] I am extremely indebted to Mr Edward Hald for allowing me to publish the facts and figures above.

use. At the Swedish Exhibition in London in 1931 there was unanimous enthusiasm for the table-glass shown.

At last some English manufacturers realized that there was a public for such objects, and the two firms mentioned at the beginning of this section started production of goods designed to compete with those of Sweden. In 1931 the agreement mentioned above was signed between one of these firms (which up till then had been addicted to the nineteenth century's lavish cutting) and the London architect. A little earlier the other firm had decided to forsake their traditional policy.

The only palpable result of this departure so far is that in a few shops one is now able to buy good modern English glass, and good foreign glass in a few more. So the situation in medium-price and expensive ware is far less satisfactory than that existing in cheap goods—a remarkable exception to the usual state of things, instructive enough to return to later. My questionnaire at the Midlands Exhibition also brought out clearly how much remains to be done. A heavily cut crystal vase was the favourite, and not the simple modern vases designed by the London architect. These obtained altogether twenty-six affirmatives, but also twenty negative ("I like least...") votes.

One is led to the same conclusion by the sales figures of the firm employing the London architect. I was told that it was some time before the new designs paid at all. Since 1933–4 success has apparently been attained and the sales of their most popular cut service in modern and in traditional design were already, in 1934, as 5 : 6. That it took so long to find a market for the new style is, according to the manufacturer as well as the architect, due less to the public than to the average shopkeeper, who lives still in the pre-Swedish era of taste. Exactly the same point was made to me by another manufacturer. He said that even some of the shops selling Swedish products would refuse to show similar designs of British origin. Be that as it may, his experience after a year was that it took him the same amount of time and trouble to sell one piece of modern design as it did to sell sixty pieces of conventional crystal. So he decided to abandon the experiment. The same manufacturer told me of some more general difficulties in the glass industry. Owing to the small volume of trade, manufacturers cannot afford to maintain well-trained staff engineers on the premises. On the other hand, an extension of the trade seems unlikely, because the public appears less and less inclined to spend money on table-glass. A young couple may buy a cheerful looking dinner-set and a tea-set as soon as possible, but very few people think of glass as an object of decorative art in the house. One may regret this, or—thinking of the Woolworth department—not regret it so much; it remains a fact that in an ordinary glass and china shop four-fifths or even nine-tenths of the total sales are of pottery.

A few more details about the firms visited. I could gain access to four only, which may detract from the value of my conclusions. Two well-known Midland manufacturers of crystal declined to be interviewed. Of the four factories seen, one employed 1200 workers, but produced mostly rolled and pressed glass, i.e. glass which is shaped by modern mass-production methods without any skill being required in manufacture. For although the creative work in Venetian as well as in Swedish or Bohemian or English glass is done by designers and not by craftsmen, the free Venetian blowing, the Bohemian or Belgian or English cutting and engraving, and even the blowing into moulds, all require considerable skill and experience. Rolled glass is poured out on big slabs and shaped by engraved rollers passing over it; pressed glass is poured into a mould and shaped by the pressure of a plunger. In firms specializing in these new and cheap processes, artistic design comes usually only into a few side-lines. In the works which I saw, I was concerned especially with two processes: the production of ornamental panels in opaque glass for the tops of shop windows, and the production of cheap glass dishes and bowls. The latter had only recently been introduced, a modern architect being put in charge of the designing; in the opaque panels modern or modernistic design cannot yet compete with the traditional patterns. Sales figures for these were given as 10 per cent. modern to 90 per cent. traditional.

The firm which, by the work of the London architect, has become known as the chief promoter of exemplary modern glass design in this country is a producer of crystal on a large scale (about 400 employees). Each year 500 new designs are needed. Of these, 150 are contributed by the architect, who is paid for two months' work but does not get royalties; the other designs—some modern, some period—are produced by the manufacturer himself, by a sales manager and by foremen in various departments. In the small high-class London works, the creative designing is mainly done by two of the directors. Both are keenly interested in modern industrial art abroad and in England, and at the same time proud of representing a firm the fame of which goes back to the early days of the Arts and Crafts Movement. The feeling of a moral responsibility has never ceased, and once more this attitude is responsible for the high artistic value of the products.

SILVER AND ELECTRO-PLATE

The following account of the conditions of design in the English silver and plate trade is very incomplete. I saw only six firms in London and Birmingham, and none at Sheffield. Only one of the manufacturers whom I met produces cheap articles on a large scale. Cutlery could not be included. However, since the Board of Trade Council for Art and Industry is preparing a report on the silver trade and will, I understand, be shortly ready to publish it, this gap will soon be filled. Readers are referred to that official document which will be based on evidence far more comprehensive than any that I could collect.

I visited two high-class silver firms that do not produce any plated articles. One of them employed about 25, one about 50 employees. One large Birmingham firm producing expensive as well as cheaper silver objects and enamel goods gave me 500 as the number of their employees. A plate manufacturer whom I saw gave work to 15 or 20 hands. Of about the same size was another silver firm to which I had access. In addition I interviewed two independent craftsmen designing for silver and plate. Units in the trade are as a rule small. Of 253 London firms of jewellers, silversmiths and clockmakers, only 22 in 1930 employed 50 or more workers.[1] Very little machinery is needed, less than in any of the industries with which we have dealt so far. As for high-class silverware, the only mechanical process used is spinning on the lathe. William Morris and his followers were strongly against this technique as being impersonal; but they could not succeed in holding up an inevitable development. To-day hand-beating has become extremely rare, and no reduction of aesthetic value can be attributed to this change. Except for the process of spinning, most expensive silver goods are still hand made. Unskilled workers cannot be employed. This is different

[1] *The New Survey of London Life and Labour*, vol. viii, pp. 147ff., 1934.

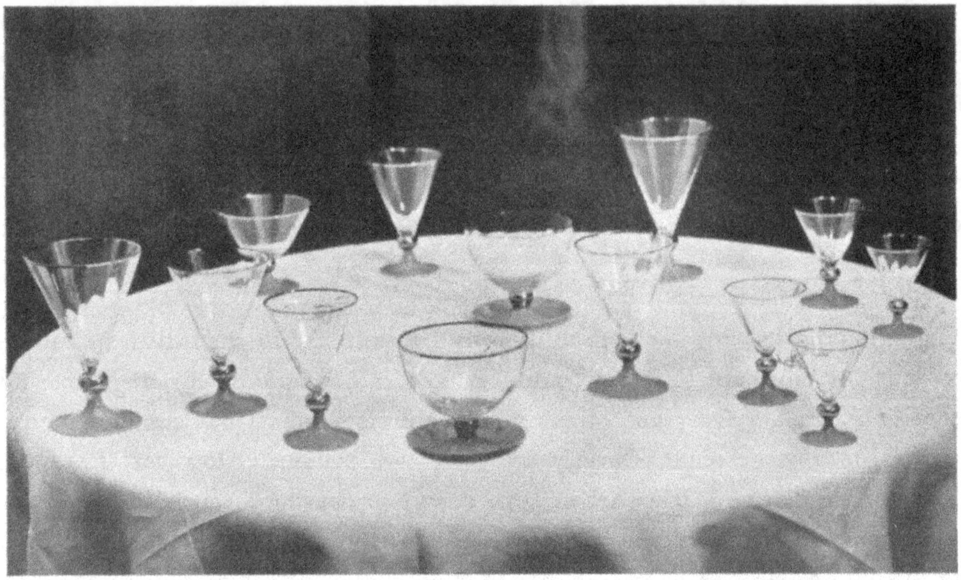

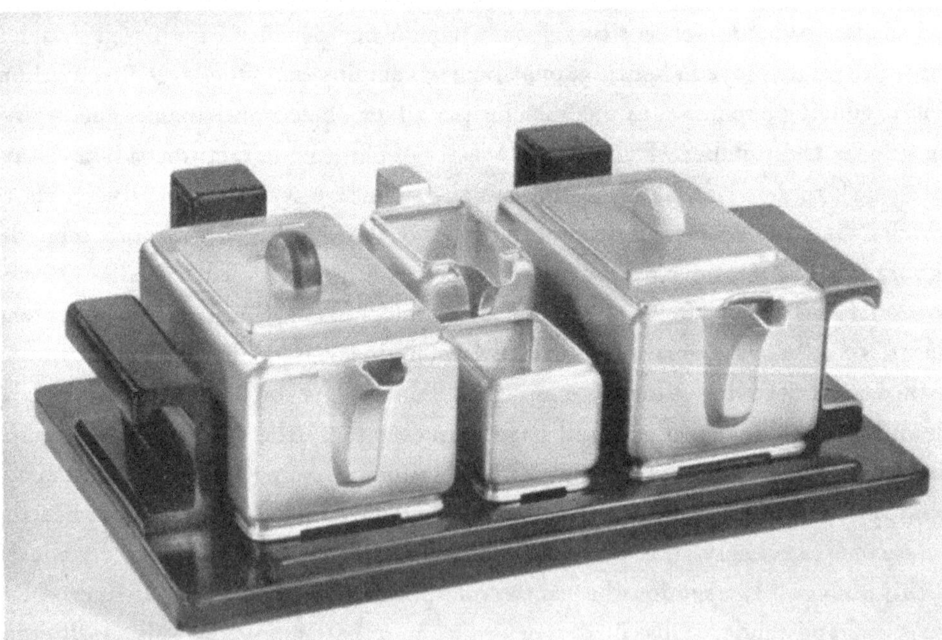

XVI. *Table glass at Woolworth's.*
Tea set designed by Harold Stabler and made in electro-plate or silver by Adie Bros. The pieces are fitted on a special bakelite tray so that they cannot rock.

in the cheap silver and plate trades, where power-presses are used for the stamping of shapes. The costs of tools are relatively high—about £30 or £50 for a cup or a sauce-boat. On the other hand, stamping is a process so much quicker and less expensive that the complete article is in the end considerably cheaper than anything hand-made. Hence the use of the power-press is advisable only where production in large quantities is possible. Of a high-class silver tea-set 6 or 12 a year would be an average sale, 50 extraordinarily good. Lower-price goods in silver may be turned out by hundreds and sometimes even by thousands. In cutlery, naturally, higher figures are reached.

The taste prevailing in English silver and plate is traditional. Modern forms were introduced about five or six years ago, both in expensive and cheaper goods. Fashions change faster now than before the crisis of 1930. Then a successful pattern might have lasted for five years and more. To-day the wholesale merchant and shopkeeper ask for something new every year. But new does not always mean modern. The public, I was told, is particularly slow at taking to the modern style in silver. You change your curtains, rugs or pottery more frequently than your silver, and certainly you will not scrap your silver because it is no longer quite up-to-date in design. It is for this reason above all that the modern style (and "modernism") make only slow progress in the silver trade.

As things are at the present moment, the proportion between period and modern forms is, broadly speaking, about 90 to 10, though in the shops of some better-class jewellers it may recently have changed to 75 to 25. or 65 to 35. We have seen in the pottery and glass trades that the modern style is absent in expensive porcelain and crystal articles though usual in cheap goods. This does not apply to the silver and plate trades. Here the less costly material is usually regarded still as a mere substitute for the more precious metal which it was once meant to imitate. No independent style has yet been developed that expresses the character of our time in terms of electro-plate.

Not that what is called modern silver is not becoming more popular on the British market, above all during the last three years. But most of these modern designs are of extremely doubtful aesthetic value. As a rule the term "modern" is applied to things with straight instead of rounded surfaces and some rather freakish angular details, resembling those which have already been described in connection with door furniture. Still it is unlikely that the infection of pseudo-cubism will go so far in silver as in some other materials. The surviving tradition is too strong for that. Indeed some English eighteenth-century silver pots, plates, dishes and jugs and some English eighteenth-century cutlery are so immaculate and simple that one does not see how anybody could improve on them. Reproductions of such pieces are on show in most shops and are more pleasing than anything modern offered by the same jewellers. Completely contemporary silver of a simple and well-balanced design is very rare in England, equally rare, according to what experts told me, in the United States, and not frequent on the Continent either. It is more often found amongst cocktail-sets, such as those designed by Keith Murray in England and Russel Wright in America, than amongst tea-sets, and more often amongst tea-sets than amongst sporting trophies. These are probably the worst products in the trade, though (or because) they are particularly elaborate. Instead of the clean and harmonious lines of the eighteenth century, the popular types now exhibit hideous Victorian motives. Golf cups with four clubs serving as a support can be found, tennis-rackets and foxtails modelled with painstaking exactitude and many more such atrocities. Are we entitled to draw a conclusion from this fact and to say that modern purposes favour modern design, traditional purposes traditional design, and that the desire for a display of art favours bad design?

The working out of new shapes in silver and plate in the few big firms is usually done by staff designers, in small firms by the owner or a foreman. Here most designing consists in copying from existing pieces without, or with only small, alterations. A disgraceful method, though one often used, is this: a wholesale merchant buys a good, expensive article, not in order to sell it, but in order to show it to a lower-grade producer and induce him to copy it at half price. Naturally, in such cases no creative designer is needed. I heard that in the Birmingham silver-smithing trade less than ten full-time designers are employed. Their salaries vary between £200 and £300 a year. In some of the best firms £600 or £700 may be paid. Occasionally these firms also employ a few independent craftsmen and designers. They receive fees of £10 or £20 a design. But their services are as a rule required only for individual commissions and not for sets to be produced for the ordinary market. Therefore nothing can be said as yet about the effect which modern architecture or art could have on the silver trade.

HOLLOW-WARE

To collect information about the designing of domestic hollow-ware, of pots, pans, kettles, jugs, etc., I visited four firms producing aluminium goods, and one producing enamel goods. The aluminium factories employed anything from 120 to 400 workers; one of them was supposed to be the largest in the country. The enamel factory, which supplies one of the most popular chain-stores, was also among the biggest units in the trade, employing about 800–1000 workers. I added evidence from a smaller Midland works manufacturing "beauty boxes" in tinplate, besides picnic outfits and hearth furniture, and giving work to 60–70 employees. Data about domestic objects made from synthetic resins will be discussed in the following section.

Quantities of production are surprisingly high, thanks to the cheap prices of most articles in aluminium and enamel. One manufacturer mentioned 1,500,000 pieces a year, one 450,000 pots and kettles, one 60,000 pots of one model which he called a very good seller. The best-seller in another firm was a kettle of which about 100,000 were sold in a year. The enamel firm counted their sales also by hundred thousands in each line. The cheaper a range, the higher the sales. In one works I was shown a cheap and an expensive electric kettle. Sales figures were as 1000 to 6.

Only a small number of new designs are needed within a year. Some types of goods remain almost stationary; where changes are introduced they are slow and likely to be more rapid only when trade is slack. Until ten years ago, most firms, especially in the enamel trade, produced only a few standard shapes and colours. Now housewives want a more cheerful appearance in their enamel pots and pans, and new colour-schemes are on the market.

The shapes of aluminium and enamel goods depend on the tools used for the stamping of the aluminium sheets or the steel sheets to which the liquid enamel is afterwards applied. Skilled labour plays an important part in the enamelling process, whereas the production of aluminium goods does not require any particular training. Tools for an aluminium article of any size may cost about

£15 or £20 and therefore manufacturers want their new models to be designed in such a way as to make possible the use of existing tools. The aluminium and enamel industries do not give much scope to the designer's imagination. Hardly any firms are able or willing to keep a special employee for designing. As a rule, new shapes originate, as it was put to me by the director of an aluminium works, "between myself and the foreman". Sometimes more than these two are involved in the introduction of a new line. Anybody in the factory may, as another director said, "have a brain wave". The works' manager may propose something, or the foreman in the toolmaking shop, or a usable idea may be found in the suggestion-box, put there by one of the work girls.

As to the aesthetic results, it seems to me that they depend on the interest taken in the matter by the manufacturer himself. In the enamel firm that I saw the managing director claimed to do most of the creative work, and his products, on sale at very cheap prices in the chain-store mentioned, are very satisfactory indeed. In one aluminium firm, the owner, sincerely interested in design, does all the drawing of new shapes himself, and again his articles are certainly amongst the best in the trade. In the factory which produces hearth furniture and "beauty boxes", it is also the manufacturer who is responsible for the appearance of new models. Some time ago he tried using a London commercial designer of some standing, but found that his boxes were too expensive and did not sell. The bulk of the "beauty boxes" are sold at 10*s.* to 17*s.* 6*d.*; the boxes designed by the London artist could not be made to sell at less than 30*s.* Perhaps they were a little better in design, but the ones designed by the manufacturer are also modern and attractive. It is depressing to see that they have to compete with imitations of cheap and poor design which none the less, solely on account of their lower price, are preferred by the buyers of a good many departmental stores.

In the aluminium and enamel trades the dominant taste is insignificant but not bad. Many articles are acceptable in shape and proportion, only a small minority are spoilt by unjustified decoration, as e.g. the stamping of aluminium to make it look like wickerwork. In pans and kettles, the reason for the satisfactory standard prevailing is once more the fact that these objects are made for use and not for adornment; in teapots and the like, the satisfactory standard

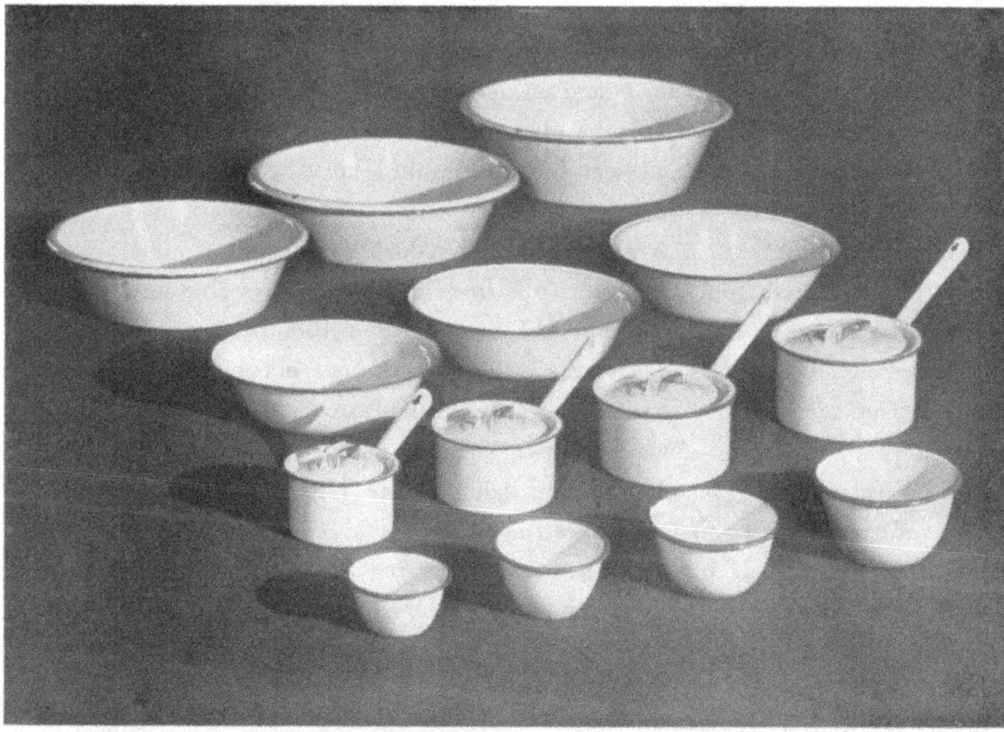

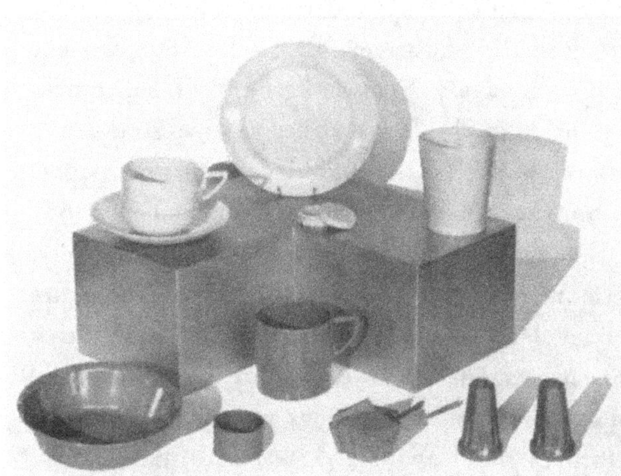

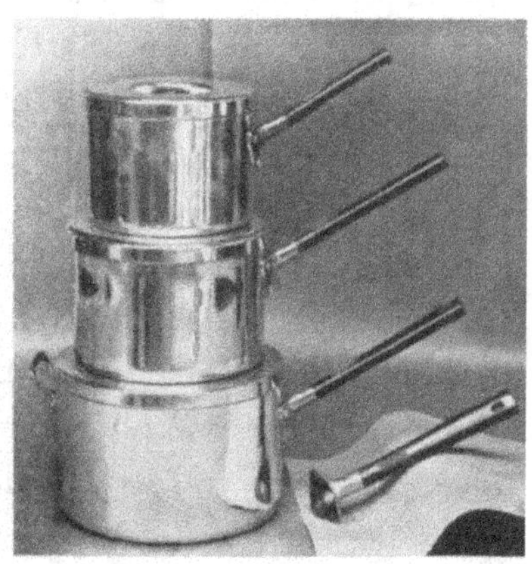

XVII. *"Pastelware" enamel goods at Woolworth's.*
"Beetleware" goods at Woolworth's.
Aluminium articles designed and manufactured by N. C. Joseph Ltd.

is mainly due to the way in which the shapes of English Georgian silver are copied. The introduction of straight lines, which is one of the features of our modern style, is in aluminium a matter of price. One manufacturer explained to me that flat surfaces can only be obtained in a stronger, and therefore more expensive, material; cheap and very light aluminium must be curved in order to become sufficiently firm.

The taste in hearth furniture is deplorable. Working through the metalware section of the British Industries Fair, one finds hardly any passable kerbs. The one manufacturer in this line whom I met did not like his own products. Of one of his best-sellers he complained that he had not been able so far "to kill the demand". Most of the hearth furniture sold is done in badly imitated period forms; "antique copper" is the favourite colour. If modern forms are introduced, jagged ornament, such as we have learnt to expect on door handles and finger plates, covers the seams and spreads over the rest of the surfaces. I find myself unable to see any redeeming feature in these products, which go so alarmingly well with most of the cheaper fireside surrounds in the new semi-detached houses.

PLASTIC MOULDINGS

Strictly speaking the term "plastics" should be applied to all materials which are shaped while in a plastic state. Usually, however, glass, clay, etc. are not included, and the term is applied exclusively to some groups of substances discovered within the last century, or even the last decades. Three groups can be distinguished: (1) cellulose materials such as celluloid (cotton and paper, treated with nitric acid and camphor) which was first manufactured in 1868; (2) casein materials used since the nineties of last century (e.g. galalite); (3) materials made from synthetic resins. The most important of these are bakelite, invented in 1908 by Dr Baekeland in the United States and made from phenol and formaldehyde, and beetle, made from urea and formaldehyde.[1] The commercial success of bakelite and the other resinous materials is entirely post-war. They were introduced into England about 1921–2, and first used for

[1] Cf. W. Landauer, "Plastics", *Trend*, vol. i, pp. 75ff., 1936. This article, published after I had finished my research, contains a more detailed account of the various manufacturing processes.

electrical appliances and the like. It was not until several years later that manufacturers realized that here were materials suitable for moulding domestic objects, cups, dishes, ashtrays.

I have visited four firms, of which one produces the powder but has no moulding plant (about 500 employees), one specializes in celluloid and turns out bakelite and beetle only as a side-line (about 1900 employees), the two remaining factories are moulders in bakelite and beetle (300, 160 employees). As in most of the recent industries only mass-production methods are used. The proportion between number of operatives and volume of output is totally different from that in old trades such as jewellery, silversmithing and pottery. To produce objects from synthetic resins, tools are made, the powder is poured into the mould and the article shaped under heat and high pressure. Even for a small undecorated ashtray, the cost of the tools would be about £20. For more complicated products it is much higher, £100 and more. Consequently large quantities have to be sold in order to make manufacturing profitable, not less than 10,000 of an ashtray, a cup or a wireless cabinet. A tumbler in beetle would cost 1*s.* 6*d.* retail, if only 5000 a year were made. You buy the same article at Woolworth's for 6*d.*, because the production comes to about 300,000 a year. Of the beetle salt-cellars at Woolworth's I heard that about 1,000,000 were sold in 1934.

On the whole the dominant taste is not bad. In simple objects for everyday use, such as the cups, tumblers, etc. at Woolworth's, it is very satisfactory indeed. Imitations of period ornament are rare, at least in England. Modernistic atrocities have infected such objects as kerbs, owing perhaps to the taste of builders or builders' merchants whom we found responsible for the jazz patterns of door furniture. Pseudo-cubistic shapes predominate in the more expensive celluloid articles. A modern-minded designer whom I met laid the blame for that on what he called "that wretched womens' dressing-table mentality". Among the best-sellers of one firm which I visited is a powder box made to look like a grand piano.

The majority of goods in bakelite and beetle are simple in outline and surface, because elaborate forms entail a more complicated moulding process. On the other hand, I was told that a flat surface, especially a vertical one, is less easy to mould than a curve, and that a completely plain surface is likely to show little ripples or blemishes which some ornament would conceal. As to favourite colours, pastel shades are prevalent, and most of the plain unmottled tints are quite pleasant. Mottling is done partly because a self-colour surface, at least in bakelite, makes moulding more difficult, partly again because manufacturers believe that the public prefers it. Larger objects are made in mottled brown to resemble wood, and it is a fact that in the case of the most popular wireless sets in bakelite three-fourths of the customers decide in favour of the brown and only one-fourth in favour of the plain black cabinets. In this respect the plastics industry seems to be in a still earlier stage of artistic development than other new trades. Steel pillars in railway stations were shaped like columns during the nineteenth century, motor-cars like carriages until about 1910. Later they freed themselves from mere imitation. The plastics industry has not yet achieved this, owing to an evident lack of artistic enterprise which I noticed in the trade.

Those manufacturers whom I met can hardly be regarded as true judges of design. One of them admired the higher art-value of bakelite goods in Germany, meaning thereby a more accomplished mimicking of oak. One wanted to prove his interest in the artistic side of the business by proudly displaying his newest jazz pattern kerbs. A third told me the story of how they got hold of an artistic designer. He once went with a customer to the station by car. On the way he mentioned that they might try employing some sort of designer. The chauffeur overheard this conversation and said that his nephew was attending classes at an art school and could do the work. "So we got him along." Although this firm now has a full-time designer, the general rule in the trade is to leave the creative work to whoever happens to take an interest in it. There may be a draughtsman on the staff who does it, or a furniture store may supply a kerb in wood and say they want something like that in bakelite, a stationer's shop may show sketches of inkstands, etc. As a rule such sketches cannot be carried out without alterations, because to design for the plastics trade involves a considerable knowledge of moulding processes. It is no use designing a nicely shaped object if tools cannot be made for it, or if in the end it cannot be extracted from the mould without considerable loss of time. This is what makes manufacturers averse to any experiments with free-lance artists and architects. One firm had arranged some competitions, but found the designs that came in to be quite useless.

Since only a few new designs a year are required—up to twenty or twenty-five at the utmost—not much money is spent on them. Thus most manufacturers muddle on with a mediocre draughtsman who may be trained as a toolmaker, and gets his £200 or up to £400 a year for adapting customers' designs or designs produced in other materials. In fact some of the best shapes to be found on the market are copied from eighteenth-century pottery or silver, the simple forms of which appear absolutely up-to-date.

I know of only one firm which has seriously attempted to raise the standard of design in its new products by appointing a young artist who had gone through the Design Course at the Royal College of Art. This firm carries on with its former lines, often totally wanting in aesthetic value (still about 75 per cent. of the output at the time of my visit), but gives the designer a relatively free hand in creating new lines and improving on old.

If this spirit were to be found amongst more manufacturers, no doubt the plastic industry could become one of the best trades in the country. Its great possibilities for the future are evident. Would it not be worth some risk for an adventurous manufacturer to make the artistic development of the industry keep pace with its rapid technical progress?

WIRELESS CABINETS

With regard to the problems of design, this is for various reasons one of the most instructive trades. First, the radio industry is one of the few that have not suffered from the slump of 1930–3; secondly, it is the newest industry with which I had to deal—its history begins with the opening of radio entertainment in the United States in November 1920; thirdly, it shows an unusual variety of methods for producing designs; and fourthly, it is the only trade where the best English models are, or at least appear so to me, better in design than any to be found abroad.

I received information from nine firms. Five of these were amongst the seven largest producers in the country. Not one of them employed less than 1000 workers, the biggest factory had 8000–10,000 employees. As to the manufacturers of particularly cheap sets, it proved difficult to collect any evidence from them. This, incidentally, I found also with firms producing very inexpensive articles in other trades. Only one of the works seen specializes in sets selling at five to six guineas for a "portable", the retail prices of all the others being about eight to twelve or fourteen guineas. Most of the cabinets on the market are made of wood; only a few firms have, within the last three or four years, changed over to bakelite. Almost all large factories produce their cabinets on the premises, the yearly output of the firms which I visited varying between 30,000 and 250,000 sets, while most smaller firms order them at cabinet factories. The designing in such cases is done by the furniture-maker, in all other cases by the radio firm.

The appearance of English wireless cabinets was, in its first phase, dependent on the historic styles of furniture. Georgian, Queen Anne, Tudor, in all kinds of adaptations and mixtures, were popular. Then, about 1930, radio design reached the state of recognizing its own particular nature and of wanting its modernity to be expressed in the cabinet. The immediate result of this was that design went modernistic, again borrowing forms from the furniture designer. This was the time of the flamboyant sound holes. Then, about 1932, "flamboyant" was often replaced by "cubistic", and for some time, in the cheaper and a large part of the more expensive trade, most of the cabinets were adorned with some jags and angles. But at the same time, about 1932–3, a reaction set in, and a few manufacturers began to turn out cabinets of simple, clear, well-proportioned shape. At first there were only two or three of them, then since 1935 they became more frequent, so that to-day the radio trade produces a few admirably and genuinely modern cabinets, a good many more of a sound and creditable appearance, and a considerable number mostly of cheap, but also of expensive cabinets which are spoilt by unnecessary pseudo-cubistic details. In classifying cabinet design in this way one regrets to see that, while there are certainly bad models in the high-price range, hardly any good or acceptable cabinets are to be found at the lower end of the trade.

A radio factory produces about three to eight new models a year, a very small demand for designs as compared with textiles or pottery. The following methods are in use for supplying designs: As a rule, firms have a drawing office under a head designer. His payment may be anything from £400–£600 a year. I heard also of higher salaries, but was unable to verify these reports. Designs are essentially limited by technical requirements. Usually the designer is given the ready-designed chassis and asked to sketch out a series of possible "overcoats" to this, for portable, console and radio-gramophone sets. His office then produces about twelve or more sketches, which are submitted to a first small committee, consisting only perhaps of the chief engineers. They weed out the majority and leave about five or six of the sketches. After that these five or six designs may be produced tentatively in makeshift materials and shown to a second conference, in which the sales manager has the last word. He may agree or may suggest fundamental changes; but in either case the artistic designer is only partly responsible for the final result. It hardly ever happens that the sales and production departments are so enthusiastic about a new design as to alter technical arrangements in order to preserve the appearance. I know of only one case where the inside of a cabinet was adapted to fit an attractive case. As to the artistic results of the staff-designer system in the wireless trade, these have sometimes been very satisfactory. But on the other hand most of the bad modernistic cabinets on the market are also the work of staff designers in radio or furniture firms.

A second method, used for the most part as a supplementary source of design, is to call in free-lance furniture designers. The reason for this combination of two methods is that the designs produced by the works' studio may be in danger of becoming too uniform and monotonous without some stimulus from outside. We have found the same argument in the carpet trade. But in only one case was I able to see much difference between the type of design supplied by the employee and that produced by the free-lance artist. Fees are high as compared with those in other trades. I heard of ten to fifty guineas being paid for a sketch which goes into production.

But the best wireless cabinets in England are those of Murphy and Ekco, the work of neither staff designers nor commercial artists. About six or seven years ago, Mr Frank Murphy decided that in order to get cabinets of a really distinguished (and at the same time distinguishing) character, it would be worth

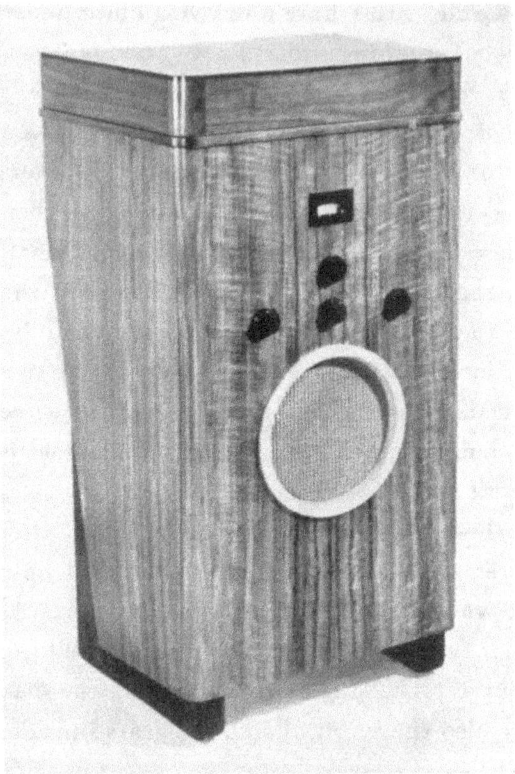
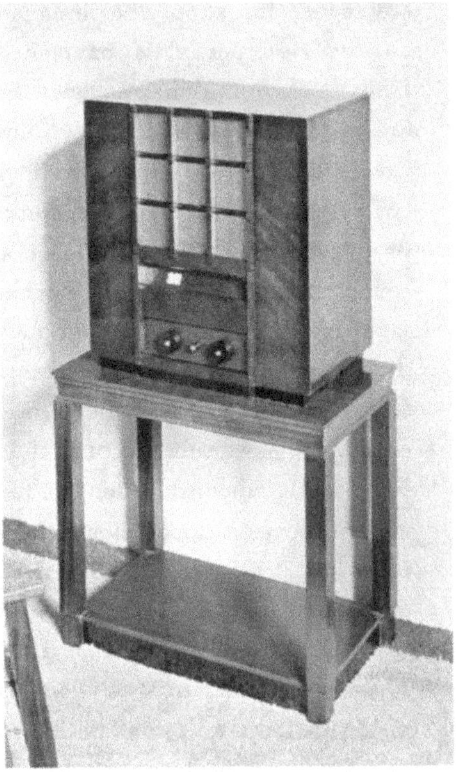
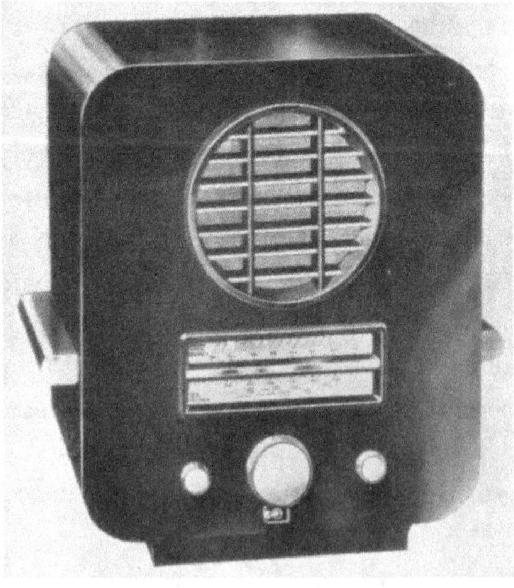
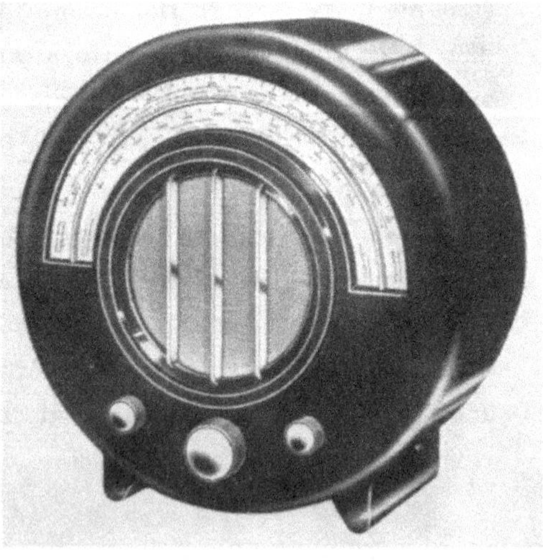

XVIII. *Murphy radio cabinets, 1936 and 1933. The cabinet work by Gordon Russell Ltd. to R. D. Russell's designs. Ekco radio cabinets, 1933 and 1934. The cabinets in bakelite, left to S. Chermayeff's, right to Wells Coates's design.*

trying one of the most uncompromising cabinet-making firms in the country. Thus he approached Gordon Russell's and the results have proved highly satisfactory in every respect. The sales figures were on the up grade when the experiment was started, they continued to increase steadily after the introduction of the new models. Only the first of these designs was particularly striking, the following designs were of a quiet, dignified style in harmony with any good furniture whether modern or old. Now the designs are no longer supplied by Gordon Russell's, but by a distinguished architect of the younger generation who for years was connected with the Gordon Russell workshops.

The Ekco cabinets are totally different in character; their story is also worth telling. The firm of E. K. Cole, Ltd. started producing wireless sets in 1930. The cabinets were made of wood by a furniture firm. By 1931, they had already realized the great possibilities that might lie in the use of those new synthetic materials which made it possible to produce the whole of a cabinet in one large moulding. For a time they bought their bakelite cabinets in Germany. When new import duties were introduced in 1932, a bakelite plant was bought and production at home started. But a bakelite plant for such big objects is exceedingly expensive, tools for one model cost about £1000, and thus the directors at the end of 1932 were not satisfied with the economic situation of the firm. It was then that Serge Chermayeff and Raymond McGrath were approached to design new models, regardless of any conventional forms already on the market. Each of them produced six sketches for which they received adequate fees. One model only was chosen for production. It was one of Chermayeff's, and his fee was increased by another £50. The shape of it was something completely new, nothing comparable existed, either in England or abroad. It was the result of a careful study of function and a genuinely artistic imagination. The firm in thus breaking away from all traditions took a great risk, and some opposition within the firm had to be met. If sales did not exceed 10,000 sets in the first year, it would be a serious failure. What happened remains as an immense credit to the English public. The new cabinet, so uncompromisingly functional in appearance, became exceedingly popular almost from the first month. About 100 per cent. more sets were sold in the 1933–4 season than in the previous season.

Greatly satisfied with this experience, the firm called in for their 1934 production Chermayeff, R. McGrath and Wells Coates. The choice fell on two of Wells Coates's designs, of which the famous circular one was produced unaltered, whilst the oblong one underwent some changes in the production department. Wells Coates's designs were even more startling and unprecedented in appearance. He approached the problem as one of designing a piece of modern machinery; whereas in the Murphy cabinets the governing idea is to create good modern pieces of furniture. Both conceptions are equally justified and the public has shown its appreciation of both, although, by preferring the mottled brown imitation-wood finish to the cleaner black one, it has, as I said in a previous section, proved that it is not yet willing to go quite so far towards functionalism as the architects are. But while the more conventional colour remained victorious, the new shape of the 1934–5 models was once more a complete success. Sales were up again by 25 per cent. on those of 1933–4.

For the 1935–6 season the same system was used. The turnover of the firm is now £1,250,000 as against £200,000 in 1930.[1] This time, sketches were commissioned from four architects. The total spent on designs came to £400, a sum unusually high in England, although still negligibly small when compared with what famous designers are accustomed to receive in America. In fact these payments for artistic value are not higher than a firm's ordinary salary to a staff designer, and the great volume of the radio trade should make it easy (and advisable) to spend such sums and more for improvements of appearance.[2]

What is the lesson to be derived from all this? First, that new and outstanding design can be a decisive stimulus to sales. How much of the success was due to the beauty of the cabinets and how much to the sheer novelty of their appearance, it is impossible to say. For such unwonted forms are bound to act as an advertisement. Chermayeff's and Wells Coates's models are conspicuous even in the worst shop window, and every child recognizes them as Ekco sets. The commercial value of this can hardly be overestimated. I realize that this

[1] Cf. *Trend*, vol. i, p. 75, 1936.

[2] Once an English radio firm bought a design or a group of designs for more than £1000 from a distinguished London firm of high-class cabinet-designers and makers. But I am not authorised to publish any statement about the success of their experiment. In any case it has remained unique up till now.

may encourage firms of a lower business morality to outdo competitors by an unparalleled and sensational showiness. But the case of the Ekco cabinets may also prove that to get something really new and convincing one has to go to the best men, and one has to pay them accordingly. This is the strongest confutation of what so many manufacturers report, namely that they need something "modernistic", i.e. as a rule some pseudo-cubistic ornament, in order to make a wireless set popular. The words of the director of one firm were that cheap sets must be "less quiet and subdued" than higher-price cabinets, in order to be satisfactory sellers. But so far no experiment of the Murphy or Ekco type has been tried by any of the manufacturers of lower priced sets. It is to be hoped that one of them will have the same courage and take the risk. The 50 or 100 guineas for design would certainly not be of any commercial importance to them either. Such an experiment would be far more convincing than any Trade Enquiries can be. As to these, more has been done in the radio trade than in any other, owing partly to the enterprise of the trade and partly to its close relation with the United States, where the use of industrial psychology for business' sake has become almost a matter of course. I know of questionnaires which were sent by an American radio firm to 16,000 dealers and 10,000 consumers, and of Home Enquiries on a smaller scale carried out by the same American firm and by an English manufacturer who entrusted the job to their advertising agents.

Such questionnaires referring exclusively to the appearance of the cabinets are of particular importance in the wireless trade, because ordinary sales statistics can never distinguish between the popularity of a receiver and that of a cabinet. Sound-reception is so essential that a customer will hardly desist from a brand which he likes or regards as especially good, just because he has some objections to the appearance of the cabinet. Also there seem to be very few people who would be prepared to pay a guinea more for a cabinet the design of which they prefer.[1] Such considerations make it almost impossible to come to any conclusions as to the popular taste in wireless cabinets solely from sales figures.

[1] The black circular Ekco cabinet was sold at 10*s*. 6*d*. more than the brown. It is probable that this partly accounts for the lower sales.

JEWELLERY

Throughout the nineteenth century, Birmingham has been the centre of the English jewellery trade. About 1820, the workshops of the goldsmiths and jewellers were still in the city, in Colmore Row, Great Charles Street and Newhall Street, while most of the workmen lived in the newer quarters to the north. The growth of the trade enabled more and more workers to set up independent firms of their own. The new firms were located in the houses of the former workers in Frederick Street, Vittoria Street, Vise Street, etc. First only one room would be converted into a workshop, later the whole house would be adapted and fitted up as offices, while the hand work was done in a shed built at the back, where a little garden had been before. Such is the origin and the appearance of most of the firms constituting the jewellery trade in Birmingham, and many an old manufacturer may well remember having played as a child in the room which is now his showroom or warehouse.

The continuous expansion of the trade went on until the war. In 1861 there were about 7000 workers in the Midland jewellery and plate trades, in 1891 there were about 15,000, in 1921 over 30,000. Since then things have changed, and jewellers complain of an unceasing decline during the last fifteen years.

To-day there are more than 300 jewellery firms in Birmingham. The majority of these employ between 10 and 30 people. Little machinery is needed, and even the power-presses used for cheap production are not so expensive that small manufacturers cannot afford to buy them. I saw eight firms of which two had more than 100, two less than 50 employees. Beyond these eight I also visited one famous factory producing watch-cases.

In the jewellery trade conditions of production and design vary considerably according to the class of work done. One has to distinguish between five types: high-quality diamond-setting, medium-price gold and gem jewellery, cheap rolled gold, silver and chromium jewellery with paste instead of genuine precious stones, the manufacture of chains, and the manufacture of enamel goods. I saw representatives of all these classes.

In expensive diamond-setting the whole process of production is still hand work.

Highly skilled craftsmen are indispensable. Commissions as a rule are given for individual pieces. No batch production exists. Often the stones are provided by the customer, who goes to a shop in order to get a jewel remodelled. Then the shop-owner—who usually has no designer on the premises—asks the manufacturer to get some suitable designs and submit them to the customer.

The prevalent taste is still entirely dependent upon the Rue de la Paix. In Bond Street some of the most famous houses have French names. Tradition rules undisputed. The wealthy classes for whom this trade caters are particularly conventional in their taste. Most of the customers, I was told, belong to the older generation. Younger people of ample means—we have found the same state of affairs in the case of high-class porcelain—prefer spending their money on something more up-to-date in character.

One can even go further and state it as a fact that the demand for high-price jewellery appears to be dying out. To wear glass chains and cheap imitation brooches or clips matching a particular dress and discarded with it, has become the common habit in all classes. The close relation between this development and the more general social changes altering the face of Europe is evident. In European society as it is and as it will be, there is no place for a jewellery trade of Victorian grandeur. Manufacturers, so far as they survive, already complain of not being able to find suitable workers. Enterprising and skilful boys prefer to start in a more promising industry such as the motor, motor-accessories, electrical or toolmaking trades.

For the manufacture of inexpensive jewellery, tool-making and die-sinking are the only necessary forms of skilled labour. And since tools and dies are usually made by special firms, the cheap jewellers' factories are run with unskilled operatives only. Their task is to use the tools and dies for producing stampings, into which the paste is cemented. A bracelet or brooch consists of two or more such stampings assembled in a certain way. Most of the variations between different products in firms of this description are variations in the use of a limited number of existing stampings. All the time the manufacturer or designer has to invent new combinations; and this puzzle never comes to an end, since retailers ask for new samples more than once a year. Designs have to be submitted at almost the same speed as in the millinery trade.

The calculation of costs in most jewellery firms is based on the price of tools, which may be anything between £3 and £25 a stamping, or about £10 or £15 a bracelet or a brooch. Firms are compelled to carry a very wide range of samples (a thousand, I was told by one manufacturer), from which the wholesaler chooses some ten or twenty and orders half a dozen or a dozen pieces of each. This is the situation everywhere except in the new factories specializing in the cheapest mass-production articles, where the organization is based on the demand of the chain stores. Here quantities like 150,000 or even 200,000 of one pattern of a cuff-link, 25,000 or up to 75,000 of one paste clip at *6d.* or 1,000–15,000 brooches at *5s.* may be produced during a year.

The consequences are the same as those already found with cheap mass production in other trades. Limitation of designs, and refusal to take any risk such as might be involved in the introduction of unprecedented shapes and patterns. This is true of England as well as of the large mass-production factories of cheap jewellery at Pforzheim in Germany which, in fact, are even more specialized than their British competitors. There is, however, one interesting and significant exception: some of the bangles and similar less adorned objects in the cheapest multiple stores are much more genuine and sound in design than any brooches or bracelets in the same type of shops. This one has to keep in mind.

So much for expensive and cheap jewellery. The conditions in medium-price goods are no more favourable to design; and here the fault lies with the manufacturers, whose attitude may be due to general narrowness of outlook or to fear—a fear easily understandable in men who have little capital for experiments. Their only capital as a rule is their stampings. Any sudden change of style would make these worthless and call for a vast set of new tools. Which of the Birmingham workshops could afford that, while struggling against the strongest Continental competition coming from the Pforzheim factory as well as the Bohemian home-workers? Thus things are left as they are and, pernicious as this policy may prove in the end, thus they prefer to muddle on. The economic situation in the jewellery trade is not enviable; that is evident. One cannot expect to find, in such circumstances, much care for the designer. It was several times stated to me as a fact that in the whole Birmingham trade no more than five or six firms keep a full-time designer. And scarcely one of these earns more than £450 a year. £300 is probably the average payment of a full-time designer. But what should induce a thrifty manufacturer to pay higher salaries, if he needs his man only for doing the jigsaw puzzle of rearranging existing stampings, and for copying from shops, trade magazines and perhaps a privately printed French or German trade quarterly to which the firm has subscribed? All this can easily be done by the owner, or a foreman, or a girl paid only £100 or £150 a year. Conditions being as they are, there is no scope for commercial artists either. Manufacturers know—as do the manufacturers of carpets—how bad are the results of such a haphazard system. One of them said to me: "We do not pretend to do original work", and another answered neatly when I asked how the designs for his products originated: "By copying and accident."

To come to my personal conclusions, the artistic standard of jewellery appears to be very low indeed. There are not many trades that after investigation leave one quite so depressed. But while in carpets the disease is modernism, in jewellery it is sheer stagnation, the complete lack of enterprise and adventure. The only change which was pointed out to me as having occurred during the last seven or eight years was the introduction of the "baguette", a gem or imitation gem of narrow, rectangular shape with sharp edges; and even this did not become popular in England until a vogue could be started in the more expensive trade by exploiting the fact that a member of the Royal Family wore a ring of the new type. Neither recent British Industries Fairs, nor the Royal Academy Exhibition of 1935 have contained much matter for excitement.

It is not easy to decide whose fault this is. Not the Exhibition's certainly, nor the Fair's. According to my knowledge of the trade, they could not have obtained more well-designed objects than the few they had. It may, if we go to the root of the trouble, not even be the retailer's fault. One case at least was reported to me, where the buyers of a certain type of enamel powder box for a chain store ordered a dozen of one shape without even asking to see the various sprayed patterns with which the boxes were decorated. I enquired whether there was anybody who *did* insist on the type of jazz patterns that is to be seen on most powder boxes and cigarette cases—the public perhaps? But I was told that this is not so. On the contrary, the firm in question had produced a small and cheap box in one plain pastel colour with only the clasp in another equally agreeable shade, and had obtained excellent sales. Later they thought a little change might be advisable and introduced some angular embellishments, with the result that the sales went neither up nor down but remained on the same level. That is but one accidental example, though a highly instructive one, and I imagine that the lesson to be derived from it would hold good in many other branches of the cheaper jewellery trade. Here lies the fault of the manufacturers. But the designer and the public are not guiltless. As for the public—even educated women with a natural feeling for what is becoming, who choose their frocks and gowns with care and discrimination do not insist upon cheap clips and brooches to match the style of their dresses, their entertainments, and the rhythm of their lives. No doubt the consumer could exert a pressure on the supply, in this trade as well as in any other, provided she or he were conscious of the problem.

Then perhaps designers and artists might at last create a style of jewellery really independent of past standards. Hitherto the modern movement has almost completely overlooked jewellery. There is a very understandable reason for this, which I touched upon earlier in connection with artistic metalwork. The modern movement means (thank heaven!) the superiority of architecture over the fine arts. This wholesome state of affairs, natural to the Middle Ages and still to the baroque of the seventeenth and eighteenth centuries, came to an end when the rapid growth of the industrial towns overwhelmed the architect's foresight. The new machinery ruined handicraft, and the new vanity of the painter, basking in the light of his message to the world, separated artists and public. Morris was the first to oppose this fatal development. Gropius's Bauhaus was the most advanced expression of the new doctrine which is to rule industrial art in the twentieth century.[1] The ardent desire for cleanness, directness, and precision led to the abolition of all that sickening decoration which had to the nineteenth century meant architecture and decorative art. Any surviving

[1] Cf. the first chapter of my book on the *Pioneers of the Modern Movement* (Faber & Faber, 1936).

decoration was rigidly subordinated to the simple lines and surfaces of modern building. That is clearly applicable to furniture, lighting fittings, fires and many other objects in the house. But where could jewellery come in? The wearing of jewels springs from the primitive and natural wish to excel, to attract notice and to conquer. How can those basic functions of jewellery be combined with the canons of our modern style? It is typical that chains and chain bracelets, built up from simple, precise and smooth joints, or cigarette boxes in gold, shagreen, or even sometimes enamel, are so much more satisfactory than the richer classes of jewellery. In designing a brooch the modern artist obviously feels uneasy. Unless there is a change in this state of affairs (and I do not see how there could be such a change without deep changes in the universal outlook on life), unless the best creative designers concentrate their most serious study on the reformation of taste in jewellery, and unless, on top of that, the younger generation of consumers insist on buying only what they really like—unless something of this sort occurs, there is little hope of more than occasional improvements.

This theory is strongly confirmed by the situation which I found existing in the watch-case industry. A wrist-watch is an article of everyday use apart from being an adornment of personal wear. Moreover, it is a new type of object altogether. When it was first developed no tradition stood in the way, no basic desire for showiness, no detachment from the true standards of twentieth-century life. The consequence is that design here has attained a remarkable perfection. Whatever shop you go to you will find attractively shaped watches, both for women and men. It appears ludicrous even to think of floral decoration in connection with a wrist-watch. And yet how easy it would have been to start a modernistic fashion. But what I was told at the one big watch-case factory which I visited (250 employees) was in striking contrast to my experience in so many other trades. Watches covered with paste jewellery or marcasite were never a saleable article in this country. Surely this is very odd. Here is the same public, which is supposed to insist on carpets of the lowest design and on jazz lighting fittings, suddenly becoming the true arbiter of good design in watches. In other industries such as door furniture the blame had to be taken from the public and laid on the distributor; but the distributor is as important for watches as for door handles and finger plates, most watches being sold through wholesale merchants, and not direct by the manufacturer to the public.

I see only one solution of this riddle. It is the same solution which was hinted at in a previous case. A watch is regarded primarily as a machine. People want above all to acquire a good movement. This attitude in itself is likely to protect the public from any nostalgia for an appearance of sham magnificence which would only detract from the actual value of a watch. It is different in clocks. Clocks for mantelpieces, etc. are regarded as pieces of furniture rather than as machines. Consequently the struggle between straightforward designs in metal and glass and the elaborate period or modernistic case is as yet by no means decided. Though the truly modern types are in the ascendant, I was told that in a medium-price chain store period clocks in stained woods were still far better sellers. But as this was partly a question of price, it is not quite conclusive.

As to details of the watch-case industry, the manufacturer who supplied me with information said that the trade is very small in England. Only some two or three firms are of any importance. Most of the movements and the majority of the cheap cases come from Switzerland and Germany. American firms and their British subsidiaries also play an important part. Designing, when necessary, is done in a drawing office on the premises. But little original work is needed. Much is based on foreign examples; and if changes are worked out independently, they are small and slow. This I gather is especially true of England, rather than of most Continental countries, where people are more inclined to fall in at once with new fashions. The firm of which I am speaking started to manufacture wrist-watches only ten years ago, and shifted over from round to angular shapes only five years ago. Alterations are so unimportant that it may happen that a pattern less popular two years ago comes back into favour this season.

LEATHER GOODS

This is another industry a large portion of which is localized in one small town. As Kidderminster is associated with carpets and Stoke with pottery, so Walsall has been the centre of the English leather trade for more than a hundred years. To-day some of the biggest and best equipped firms have shifted to London,

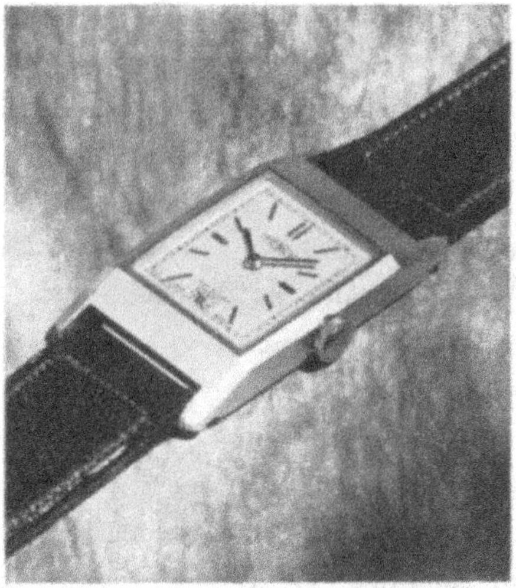

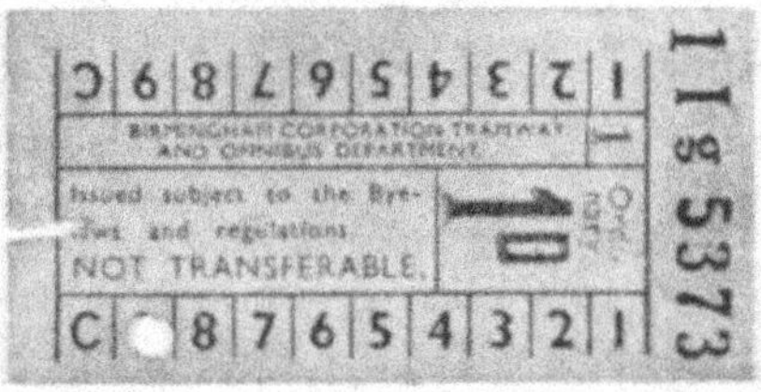

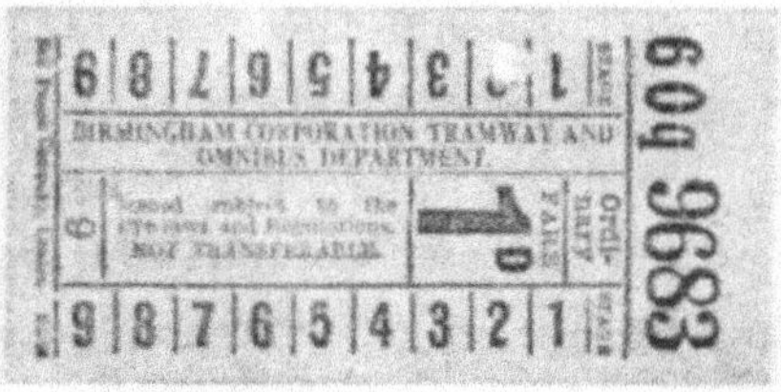

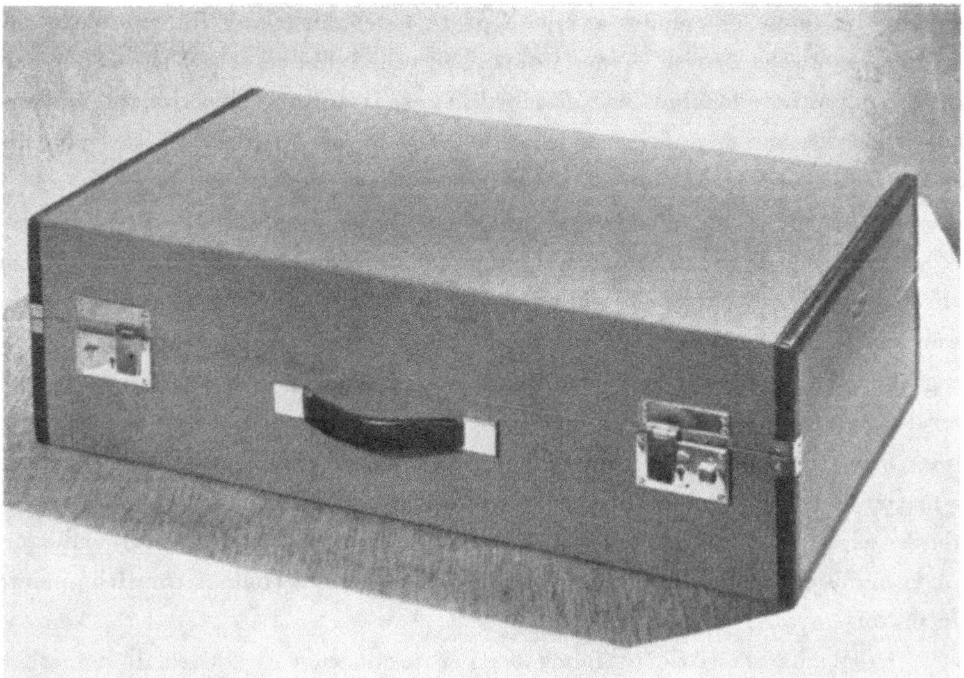

XIX. *Wrist-watch made by the Dennison Watch Company.*
Two tramway tickets showing improvements in lettering due to the municipal authorities of an English city.
Travelling case for motor-car, made by Barrow, Hepburn & Gale.

and much of the fashionable trade also prefers the atmosphere of the capital to the somewhat oppressive air of the Midland town. The vast majority of the leather goods produced at Walsall come from factories which are not limited to fancy articles. As a rule heavy leather sections are added, where belting, saddles and similar goods are manufactured. This is said to have influenced the character of Walsall fancy goods which are supposed to be heavier, more solid and more practical than the corresponding London products.

Another distinguishing feature of the leather-goods industry is that the manufacturer gets his material from the leather factory already prepared and dyed. As far as grain and colour of the skins and hides are concerned, the tanner and not the manufacturer is as a rule responsible. For more expensive articles the leather-goods manufacturer may make a careful choice among samples available, or even suggest certain patterns or colours to the tanner. For cheap things the leather-goods manufacturer has to rely on what the leather producer offers. The range of colours available is to a certain extent influenced by the suggestions of the British Colour Council, to which body I found more than once the setting of new fashions was ascribed. There is nothing objectionable in that; but if the leather manufacturer takes a fancy to jazz patterns or to embossing cheap leather so as to make it look like more expensive reptile skins, and if thereby he spoils the discrimination of the buying public, the responsibility is his, and not so much that of the small handbag manufacturer who cannot afford to resist. Perhaps one ought not to be too orthodox about what is called embossing, i.e. the process of artificial graining, by pressing the skin with an engraved metal plate. It may be a process which leads in most cases to unpleasant results, but even natural grain would not be as clearly visible as it is, had it not been brought out by the rubbing and creasing of the heated skin. It is all a matter of degree, and I certainly do not want to give the impression of pleading for senseless patterns designed not even by the leather firm but by an embossing plate manufacturer, whose rôle in the game resembles that of the litho-printer in pottery.

A third characteristic of fancy leather production is the small amount of machinery required. Besides sewing, only polishing and some marginal processes such as skiving and paring are done by machines. The cutting of the leather

which they claim to know "like their gloves". Its demands are said to be remarkably static. Though changes are continually introduced, they are never allowed to be drastic. Women buying their bags in such a shop know beforehand what they want. They will not buy exactly the same thing as they had before, but usually something very similar. All this is different in the cheapest type of multiple store. Leather goods here must be more catching. A girl wandering round and gazing at the counters very often does not know what she is going to buy, or she may take the opportunity provided by the purchase of a pan or a writing-pad to have a look round the other departments as well. If she sees a purse or pochette which strikes her fancy she may buy it and not grudge the coppers. Therefore leather articles in green, beige and red find a public here, whereas the demand of the other chain store is almost limited to black, brown and dark blue. And for the same reason, the 6*d.* store is sometimes able to go in for something more modern and adventurous. The cheapest of shops certainly cannot be reproached with stagnation.

Some more details of the popular taste in handbags may be added. During the last few years reptile skins and imitations of these have been fashionable. Forty per cent. of the orders placed by a cheap chain store consisted of imitation reptile, and 35 per cent. of the total production of a big London house was manufactured from genuine reptile. The remaining 65 per cent. in that factory were divided as to colours in the following way: 40 per cent. brown, 15 per cent. black, 15 per cent. blue, 30 per cent. all other colours. The chromium vogue has affected the leather trade as it has so many other trades. Once more I found that London leads the fashion. The North and Scotland in particular ask for bigger and heavier bags of a more solid appearance and more useful shape. Fashionable articles on the Continent seem sometimes flashy and over-smart to English observers. But in spite of that Vienna still maintains the highest standard in bags, pochettes, etc. In Germany, Offenbach is the centre. It was emphasized to me that the trade in Germany and Austria had a great advantage in that the women of those countries are so much more "bag-conscious" than in England. This is supported by the fact that there are far more leather shops in Germany, Austria and Czecho-Slovakia than there are in England.

The situation in travelling goods is different. Changes which have occurred

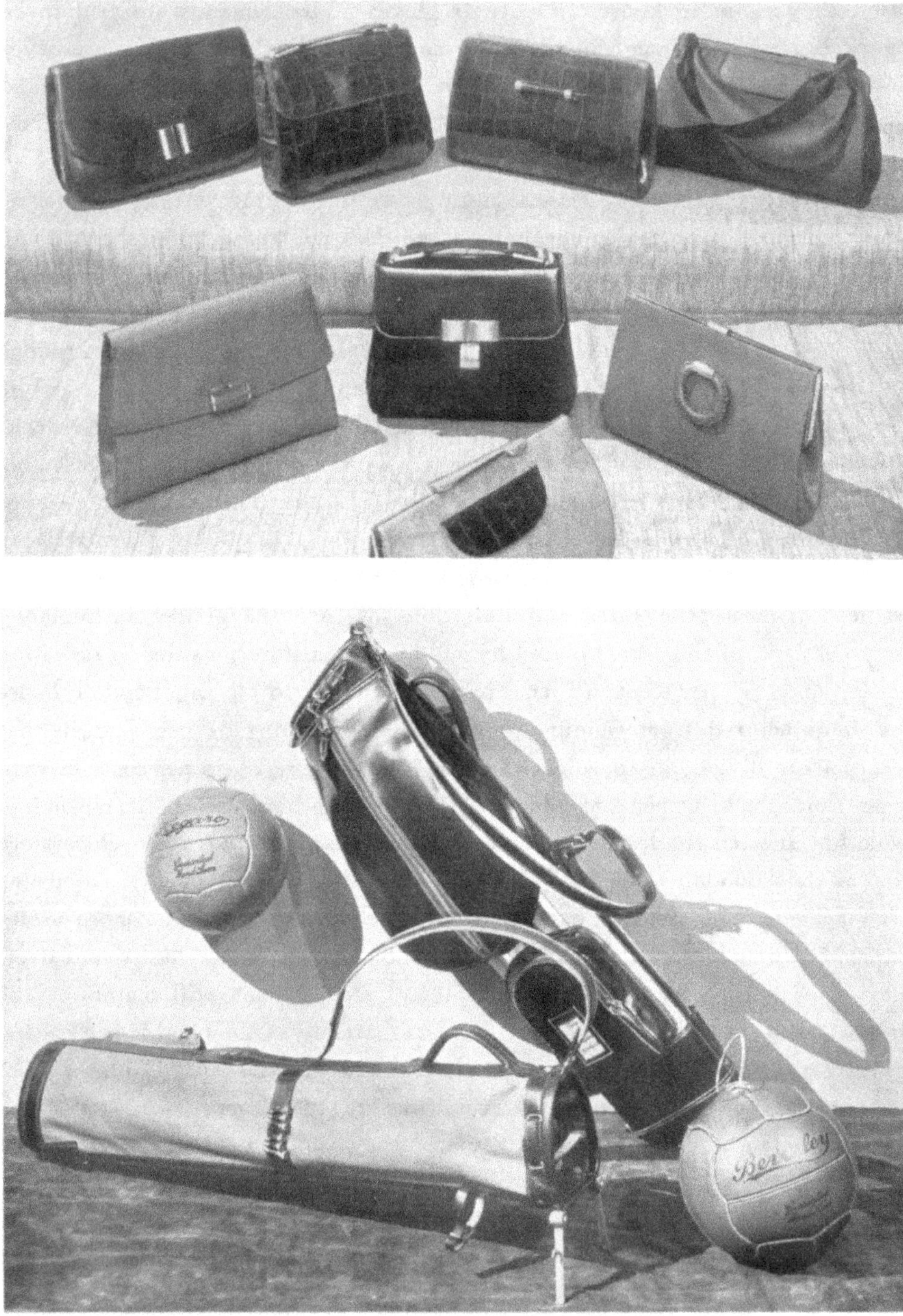

XX. *Handbags and sports articles, made by Barrow, Hepburn & Gale.*

during the post-war years have been rather in matters of use than in artistic style. Travelling habits—above all, the popularity of the motor-car—have created a demand for light models. The heavy leather travelling bag, fitted with every implement for the perfect gentleman's dressing-table, has become more and more the privilege of a very small, wealthy and conservative class, and has already something of the flavour of a museum piece. In cheaper goods rexine is supplanting leather. In trunks plywood or fibre board is used. Recent developments also are the wardrobe-trunk and the Revelation suitcase. Another new departure was the introduction of lighter colours in travelling goods. 1934 saw trunks in Marina green, 1935 in Jubilee blue. In a few months they will probably be all Coronation blue and Coronation red. But in spite of all this, the style of English travelling goods appears to remain almost stationary. And since here—as well as in sports goods made from leather—practical needs have once more dictated shapes and details, a highly satisfactory standard is maintained. In exhibitions abroad (as e.g. at Copenhagen in 1932 and at Milan in 1933) English trunks, travelling bags and sports articles are as a rule much to the credit of the English nation.

In cheap attaché cases design is mostly insignificant, but again no sham modernistic or period outrages appear. In handbags a little more art can be applied, with the result that there is greater danger of blundering. But when compared with the carpet and some other trades English fancy leather production keeps to a fairly high standard. That very cheap purses and pochettes look cheap cannot be helped. In leather more than in many other materials, such as pottery, the charm of the finished product depends on the quality of the material used, its grain and texture, its smoothness and roughness. But there is a difference between looking regrettably cheap, and looking cheap and nasty; and nasty-looking leather things are exceptional. On the other hand it seems to me that it is not the most fashionable West End firm which produces the most perfect articles. The Royal Academy Exhibition has shown that eccentricity and mannerism are more likely to be found among the smartest of handbag makers than among firms of a more practical character. Those especially English qualities which delight our eyes in English sports and travelling goods may easily be lost when designers search too exclusively after French "chic" or Austrian grace.

PACKAGING

As I mentioned at the beginning of this book, information about packaging was collected only as a subsidiary matter. Cartons, packages, boxes are on the border between the territory of the industrial designer and that of the commercial or advertising artist. I saw no more than two firms specializing in packings, of which one was Messrs Boxfoldia of Birmingham (about 350 employees), the other one of the biggest producers of bags, boxes and general stationery in the country (about 5000 employees). I can also add some evidence kindly supplied by Messrs Cadbury Bros., for whose products wrappings and boxes are naturally of great importance.

In the specialized firms I was told that most of the packages are produced to their own designs. Only rarely have customers decided preferences, and where they have it is usually for the reproduction of a successful old design. Such a design may be exceedingly ugly, or even make the name of product and producer completely illegible, yet the order would be carried out and no attempt at improvement made. One of the manufacturers told me that he makes it a principle not to educate his customers, though the temptation is sometimes great to prove to the client a practicable increase in clarity and appeal by the scientific methods of industrial psychology.

All the designing is done by members of the staff. In the larger of the two firms visited there are more than 100 draughtsmen working. The head designer receives a salary of about £800. Free-lance artists may also submit sketches. These are liked by some firms because of their "un-tradey" appearance, whereas other firms regard them as useless on account of their lack of insight into the requirements of colour-printing, etc. Prices for sketches vary between 5*s.* and £20, the usual price being about two to four guineas. The staff designers are often sent to art schools (sometimes at the firm's expense), to educate their taste and keep it alive. Payment of draughtsmen is very low; a journeyman lithographer earns more than a draughtsman, a preposterous state of affairs. The atmosphere in the smaller of the two factories seen appeared to be cheerful and lively, with a director in favour of a good modern style and designers able to do attractive, simple, up-to-date work—in the rare instances where demand for such articles occurs. In the bigger works the director was also proud of his productions, although the head designer realized that 90 per cent. of the designs which he had to produce were devoid of artistic merit. In display boxes the case was different. Here a great deal of technical ingenuity is applied to the invention of new shapes which can be cut from one sheet of paper and yet open for display in an unexpected and attractive manner.

The general experience of Messrs Cadbury's with regard to the popular appeal of good design in wrappings and boxes is wide and depressing. They would certainly like their packings to be in harmony with their clean, simple and bright factory buildings, and with their epoch-making garden city, but they know that they must act slowly and step by step.

A distinction must be drawn between the wrapping of ordinary chocolate bars, and the boxes for assorted chocolates. With the cheaper lines one cannot be too cautious about change. In spite of this, new and much better lettering has been introduced, and completely new wrappings replace the old ones by degrees. These are doubtless more modern, but it is a pity that the dominant zigzag motive recalls the modernistic tendency of the modern style rather than the sounder and more genuinely modern trend. All designs are evolved by the factory staff, i.e. girls who are trained in an art school, or discovered by competitions open to everybody in the factory.

Of special interest for our purpose is the question of Christmas boxes. Here the public is offered the same goods, equal in quantity and price, but in boxes with different pictures on them, so that the customer can express his likes and dislikes more directly than in most other cases. This constitutes the importance of the Christmas boxes as a means of gauging popular taste, although it has to be admitted that the taste of the retailer who may prefer a certain type of subject matter (such as Venetian gondolas or hunting scenes, for example), and refuse others, is again not eliminated. As a rule pictures for Christmas boxes are bought from printing firms. The aesthetic quality of 99 per cent. of them is deplorable; and some of the producers know this as well as the chocolate manufacturers who use them. But there has always been a common understanding that improvements are impossible because the buying classes would not appreciate better design.

However, Messrs Cadbury felt that such "defeatism" would be out of accord with their numerous other social activities. They wanted to test the underlying truth or falsity of the common complaints as to the low artistic taste of chocolate consumers. A first test was made with admirably tabulated sales' figures showing the results obtained from the total sales of Christmas boxes, arranged according to three grades of sales (good, average, poor) and four grades of price (high, higher medium, lower medium, low). The result was disappointing. Designs of a slightly more acceptable artistic standard sold well only if used with the more expensive qualities. The next experiment was a competition held under the chairmanship of Sir John Lavery in 1933. It yielded 2000 entries, out of which one very pleasant stylized design (Masquerade) was chosen and put on the market at a price of 6*s.* 6*d.* for the 2 lb. box. Sales compared with those of other boxes of a similar price kept at a medium level, just between grades 2 and 3. This result encouraged Messrs Cadbury to risk a further experiment, this time on a larger scale. Ten well-known painters and illustrators were asked to design new Christmas packings. They were given a free hand, and their paintings were reproduced, specially advertised and put on the market.[1] The following artists took part: Ph. Connard, E. Dulac, M. Gertler, Dame Laura Knight, C. R. W. Nevinson, D. Procter, E. Procter, A. Rackham, G. Sheringham, A. Watts. The only comparable experiment which I have been able to report in this book is that carried out by Messrs Foley China and discussed on p. 77. But though twenty-seven artists worked for Foley's and only ten for Cadbury's, and though the choice at Foley's covered a wider range of tendencies and went further in considering the modern trend, the fact remains that Messrs Cadbury were, at least to my knowledge, the first in this country to arrange a competition of this kind. That deserves a special mention in the history of modern industrial art in England.

The financial result was not satisfactory. I am greatly indebted to Messrs Cadbury for giving me permission to quote the following figures. Making allowance for the advertising which the boxes received and taking the average sale of an ordinary box selling at the same price as 100, the index figures for the boxes were as follows: Rackham (best sale) 45 per cent., Gertler 35 per cent.,

[1] For illustrations cf. *Commercial Art and Industry*, vol. xv, p. 198 ff., 1933.

Watts 25 per cent., Knight (Circus) 25 per cent. The other boxes go down to 21, 20, 16, 12, 6 per cent. of the average sales. If asked to suggest a reason for this discouraging result I should say that probably the fault lay in part with retailers who did not like some of the designs and therefore did not give them adequate display. Genuine interest in modern art and snobbery—which certainly helped to make the Foley China Exhibition at Harrod's a success—in this instance could not influence sales. To most purchasers of the Cadbury boxes, names like those of Laura Knight or Mark Gertler would convey nothing. On the other hand, this may make the results all the more valuable as an unbiassed expression of taste.

Have we then to take it that the public, by refusing these boxes, has expressed its reluctance to accept art? I am personally of the opinion that such a generalization would be unjustified, and that much of this commercial failure was due to the attitude of the artists. The approach adopted by some of them to the designing of a box for a given commercial article was certainly not the right one. Neither a circus scene nor a still life of fruit has a natural and easily understandable relation to chocolate. Thus it seems to me that the chief mistake here was once more an arrangement by which design was not conceived in harmony with function but just as an afterthought. If this interpretation is right, then the failure of the experiment is not really conclusive, and it is to be hoped that Messrs Cadbury will some day decide to start another experiment by commissioning boxes not from famous painters, but from the best advertising artists in the country. I do not doubt that this method would produce a more encouraging result.

Unfortunately space does not allow me to deal with modern advertising art in England. Recent improvements here have been more rapid than in most other branches of design. England was one of the pioneers of the modern poster and of modern lettering, and though now packaging and display (above all, window display) are, on the whole, behind those of America and Central Europe, the publicity departments of the London Transport Board, of some of the Railway Companies, of Shell-Mex and Austin Reed, or some recently rebuilt shops such as Jaeger's, Simpson's, etc., or a periodical like *Shelf Appeal* (started in 1933) show how strong the desire for better commercial design has become. There are now, I am sure, at least a dozen designers in England who could easily compete with the foremost Continental and American commercial artists. It is greatly to be regretted that they do not get more chance to work for industrial art proper.

PRINTING

The production of packages is closely connected with the printing trade. I tried to complete the scanty information which I had collected about packaging by seeing some printers, though only four altogether. However, I felt entitled to regard this as sufficient, because the problems of printing are somewhat different from the main problem of this book. One may say that the difference is only superficial, the product here being the printed text, the use its action upon the human mind, and the rôle of the printer to decide the appearance of the product with an eye to its use. He must shape or choose the letters composing the text, arrange the layout and add some decoration. But there are two arguments against the comparison. For one thing the customer buying a book usually wants it for its content only. Moreover, he has no choice as to shape and appearance which he would have in acquiring a gas fire or a wireless set. In catalogues, on the other hand, the question of buying the printed product itself does not arise at all. Catalogues are meant to advertise, i.e. to induce prospective customers by artistic and psychological means to become interested in the articles illustrated. The same is true, in a wider sense, of a firm's note-paper or even of the note-paper used by a private individual.

But above all the printer's share in the appearance of his products is much smaller than the public usually thinks. As to the printing of books, it hardly ever happens that the printer is responsible for layout and type-face, this being chosen by the publisher. As to the printing of catalogues, most of the layout, decoration and choice of type is due to an advertising agent or the advertising expert on the staff of the firm giving the order. And as to note-paper, the vast majority of headings are designed by somebody attached to the die-sinking firm which provides the printer with the necessary dies. The consequence of all this is that only ten or twelve firms in the whole printing trade keep full-time designers. And even some of these find it hard to retain their services, owing to the irresistible development which is taking away from the printer more and more of the creative work. Still, as advertising art in England is, generally speaking, on a higher level than industrial design, the increasing importance of the advertising artist is certainly no matter for regret. On the contrary, the new system seems to work quite well, and certainly better than the old one which was responsible for nineteenth-century catalogues and stationery. Also advertising firms in London find it easier than do printers to get into touch with the best artists available. Some of them are said to pay extremely high fees to the artists designing posters or packings.

How small is the influence of the printing firm on the design of their products may be illustrated by some figures. In one firm that I visited 50 per cent. of the work done was based on sketches from commercial studios, and among the remaining 50 per cent. in only 1 per cent. was the printer given a free hand, the remaining 49 per cent. being dependent on customers' wishes, former catalogues, existing headings, etc. In another firm I was told that 90 per cent. of the catalogues are carried out to the designs of advertising agents, and 99 per cent. of the stationery to the dies designed and produced by die-stampers. This applies to all but the very highest class of work. In one of England's best printing firms I was told that the customer who is prepared to pay more than is just necessary, in order to obtain something as near perfection as possible, would recognize the printer and his designers as experts and leave the whole layout of a catalogue, leaflet or folder to them. But even in the firm I am referring to this is the case with only 50 per cent. of their orders, the other 50 per cent. is carried out to the customer's suggestions.

So far as taste in English printing is concerned I want to repeat that lettering is almost everywhere of an excellent standard. As I had to say when dealing with commemorative tablets, name plates and the like, even the small town or small firm is as a rule able to obtain satisfactory work. The same men to whom is due the revival of lettering in English metalwork are naturally responsible for the excellent state of lettering in English books and on English note-paper. Once more I wish to do honour to Sir Emery Walker, to Edward Johnston, and—to mention one more of the pioneers—to J. B. Fletcher and his Leicester College of Art, for what they have done to raise standards in England as well as on the Continent. But though for example in Germany, owing to English influence at the beginning of the century, things have very much improved, the average standard is still far below that existing in this country; inexpensive books are still often deplorably printed, to say nothing of the disgraceful lettering found in cheap French novels. The diffusion of well-designed type into low-price stationery and books in England is to a large extent due to the firm of W. H. Smith and to their energetic propaganda before the war. For a time they collected together in their business a number of the best English book-production artists.

In note-paper the standard usually remains good as long as headings are limited to lettering. But most business headings are decorated in some way, and very often regrettably. Looking through the hundreds of letters which I received during the period of my research, I find that only very few have really good headings, 5 per cent. I should say; one-third are definitely bad and vulgar; the rest are traditional, neither offensive nor especially attractive. This sounds depressing, but in fact it is not. We can never expect nor have we ever found a better percentage elsewhere. And considering German or French firms, the lettering in France would usually be more reactionary, and in Germany more distorted by misunderstood modernism.

In catalogues the situation in England is similar. Lettering is frequently satisfactory, the layout more often than not appalling. Only by examining the work of a very few firms can one get an idea of the improvements which undeniably have been made. But the whole of this problem is again too closely tied up with the larger question of display to be dealt with here. What I want to emphasize is that the question of artistic printing is scarcely one which nowadays concerns the printers. Only a book setting out to discuss advertising as an art could successfully go into the matter.

MOTOR-CARS

Several points about the motor-car industry are of special interest in our connection. For one thing the motor-car is a creation of the twentieth century. The first patents were taken out by Daimler and Benz in 1885 and 1886, the first cars were built in the 'nineties, the rapid growth of the trade set in shortly after 1900. There are still many people alive who can remember a time when there were no cars. I had the pleasure of meeting one of the pioneers of the English automobile trade and listening to his description of the first four-wheel petrol cars in this country, of their bodies built to emulate coaches, and of the development of car-bodies towards an independent design.

Another noteworthy fact with regard to motor-cars is that they hold a central place in public interest. Changes in their appearance are discussed everywhere and seem to concern everybody, while new types of electric fires or leather handbags steal on to the market without much notice being taken. The ambition to recognize the various makes of car in the streets has infected all boys between four and seventy. The annual Automobile Exhibitions at Olympia are, if I am not mistaken, more crowded than any other exhibitions in England (20,000 to 35,000 visitors a day). To have a car is the longing of clerk and shopgirl, to have a smarter car the longing of the wealthier classes. Consequently the trade has been spared almost all the troubles of the economic crisis. In 1921–2 Great Britain had 875,000 cars, in 1935 1,475,000. The yearly production amounts at present to over a quarter of a million. But the United States produce more than ten times as many as this (about 3,250,000 cars in 1935), and during the boom period before 1929–30 turned out four to five millions a year.

All this creates an extremely adventurous and cheerful atmosphere in the motor works. The directors and managers are genuinely proud of their buildings, their organization and their products, the staff shows one round with the attitude of a tourists' guide in Italy. Evidently such an attitude is bound to foster design.

One has to distinguish between three classes of bodies. First the all-steel body of the mass-produced car. The works in which such cars are manufactured are laid out in the most up-to-date way, with conveyor-belts, etc. Only one of these firms has a steel-pressing plant of its own, the other three order their steel bodies from three special independent works. The designing is done in the car factory and not by the pressed-steel manufacturers. The second type is to be found in medium-price vehicles. Instead of mass production we get what is now often called batch production. Bodies in this intermediate class are built up on a composite method. More wood is used and less metal, which leaves greater scope for individual skill and craftsmanship. As for the most expensive cars, the coachwork is supplied by special body-building firms. Manufacturers such as Rolls Royce and Bentley do not build or design bodywork. The same applies to some of the best-known works producing cars at about £400 to £800. They have fixed agreements with one or more coachbuilding workshops. Some of the London coachbuilders have an international reputation; but only one or two of them are able to specialize in individual commissions, the other ones usually carry out orders for 10, 20 or up to 50 standardized bodies at a time.

I went to two of the three biggest mass-production automobile factories in the country, one employing 18,000 people and producing about 100,000 cars a year, the other employing about 10,000 and producing over 50,000 cars.[1] I added one smaller mass-production factory of cheap vehicles (3600 employees, about 25,000 cars), two big firms each with a medium-price and a higher-price range (6000, 7000 employees, 12,000, 20,000 cars), two smaller works of high-class medium-price cars more or less specializing in sports models (500, 700 employees, 1000, 2500 cars), one factory turning out very expensive chassis, and three coachbuilders (150, 400, 400 employees) of whom one does hardly any special work for individual customers, the second some, but not much, the third so much that only one-third of his production is batch-orders.

As to the designing of bodies, this is as a rule not done by trained artists. A few firms employ a special artistic designer and pay him well (£600–£1000, in two or three cases up to £1500), but even they admit that he is far from free in his work. Almost everywhere the ideas for new models originate with the director or manager, and, incidentally, about eighteen months before the car is put on the market. He may make some rough sketches or more careful drawings, according to his ability, or else describe to the body engineer or the coachwork manager what he means. Only after this are more detailed drawings worked out by the designer and submitted to a board of experts (in one factory called the Elegance Committee), in which the sales manager has a particularly important voice. The next stage is usually the construction of a model, a yard or so in length. Fresh discussions take place about that, and then finally a first experimental car is ordered. The importance of the body designer's own taste in such a procedure naturally varies. His interpretation of the director's ideas

[1] These and the following figures were obtained about eighteen months or two years ago (1935).

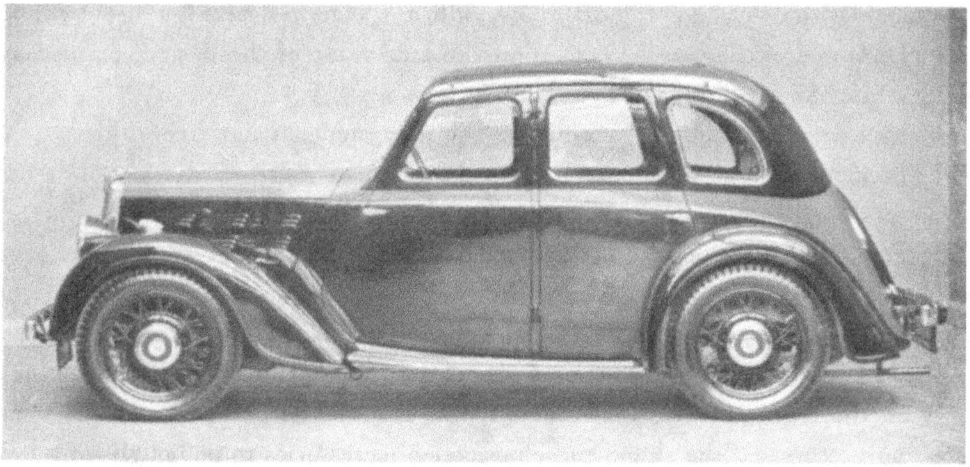

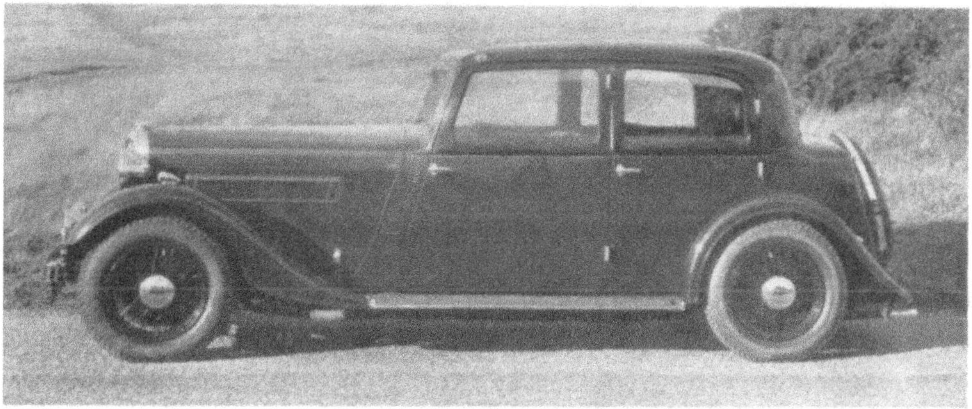

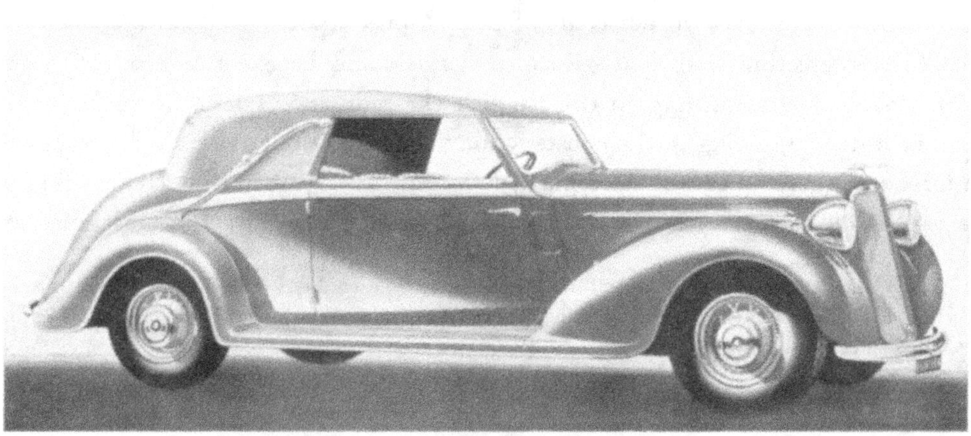

XXI. *Morris Ten-Four Saloon, 1935–6. Price £172. 10s.*
Rover Sports Saloon, 1935–6. Price £298.
Humber Eighteen Foursome Drop Head Coupé, 1935–6. Price £510.

may enable him to make out of an insignificant sketch something beautiful in proportion and outline, or he may spoil the true value of the design, or he may leave it unchanged in all its main features.

The director or manager originating the new model is not free either. Constructional considerations dominate his decisions to a large extent. To-day it seems as though equilibrium has been reached between chassis and body as factors influencing and influenced. In one firm the body and its appearance was even considered to be of such decisive importance for sales that the suggestions of the body designer had to be followed by the chassis engineer. This system, however, cannot be introduced into any mass-production firm. Here design depends not only on the requirements of the chassis, but also on the properties of the steel-pressing process and the shape and character of accessories to be bought at a low price. If the glass for the windscreen is offered more cheaply in one shape than in another the artist has to work this shape into his designs. The freest designer is of course the one working for the most expensive coachbuilder. But even here it appears that most of the creative work is done by the manager or director.

The large majority of English cars are of the saloon type, although within the last year coupés have become more popular. In an estimate which was kindly given to me by Mr E. C. M. Burrell, of the Institute of British Carriage and Automobile Manufacturers, it was said that in 1936 about 88 to 90 per cent. of the cars produced were of the saloon type. In colour, black, dark blue and grey, according to the same estimate, account for at least 80 per cent. of the bodies made. When I started my research in 1934, grey was still less favoured than black and blue which then constituted in five works questioned 48, over 50, over 50, 65, and over 70 per cent. of their total output.

More interesting than statistics on colour would be exact figures as to the prevalence of *streamlining* in the English car industry. These, however, could not be obtained, owing to the constant fluctuations of design which are going on. Moreover, there is as yet no agreed definition as to what the streamlining of a car really means.[1] Must it satisfy the definition made by the most exacting

[1] Incidentally, the word "streamlined" was already used by Frank Lloyd Wright in his lecture on the Art and Craft of the Machine, delivered in 1903 (F. L. Wright, *Modern Architecture*, p. 19, Princeton, 1931).

engineer, or can the word be applied to anything with long slanting lines and curves? I once saw something in a shop window which was called a streamlined lipstick. I also know of an inquiry addressed by a car factory to a firm of bakelite moulders for a streamlined ashtray. Should one not ostracize such a hackneyed catchword altogether? Or should one rigidly limit its use in the car industry to the rear-engine models of the Trojan, Crossley, Mercedes type, and perhaps the few tank-shaped cars such as Chrysler and Singer introduced in 1933–4? The idea of the streamlined motor-car was first conceived by engineers of vision. The earliest models with a rear engine were built by H. Rumpler in 1921 and by Sir Dennistoun Burney in 1930.[1] They were before their time, and in spite of Sir Dennistoun's passionate propaganda in England and America the trade did not take up the idea of the rear engine at once. Only hesitatingly did a few firms such as Trojan and Crossley begin to produce such cars about 1932–3. As to the more common types of streamlining, i.e. the introduction of slanting windscreens, slanting radiator fronts and outcurved backs, the United States, where the economic crisis had made manufacturers of cars eagerly design-conscious, took the lead, and from America the fashion came to Europe. On the Continent streamlining of this sort began to be applied about 1931. The first year in which the new shapes could be frequently seen in England was 1934; by 1935 there were in the whole of Europe only a very few new models which were not streamlined in the wider sense of the term. There are considerable differences in degree, but the general tendency is obvious.

If I now pass on to the artistic merits of English cars, the first thing to be said is that streamlining in itself does not increase nor reduce aesthetic value. It is necessary to emphasize this, because in a trade so rapidly developing and so vitally representative of our age, one is bound to be affected by fashion, even if one tries to keep a detached view. Pieces of furniture designed by a good architect in 1912 still seem as good and topical as they were in the year when they were first introduced. A motor-car designed before the war or even before 1930

[1] Illustrated in *The Autocar*, vol. lxv, pp. 545ff., 1930. An illustration of a streamlined car strikingly similar to the Chrysler-Singer type is to be seen in a book on Modern Architecture published as early as 1928: Peter Meyer, *Moderne Architektur und Tradition*, p. 10, Zürich, 1928.

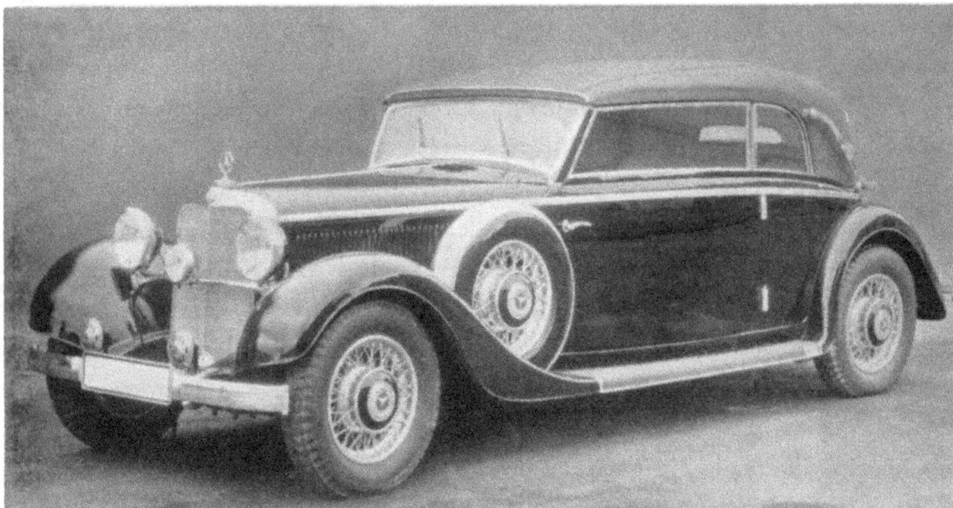

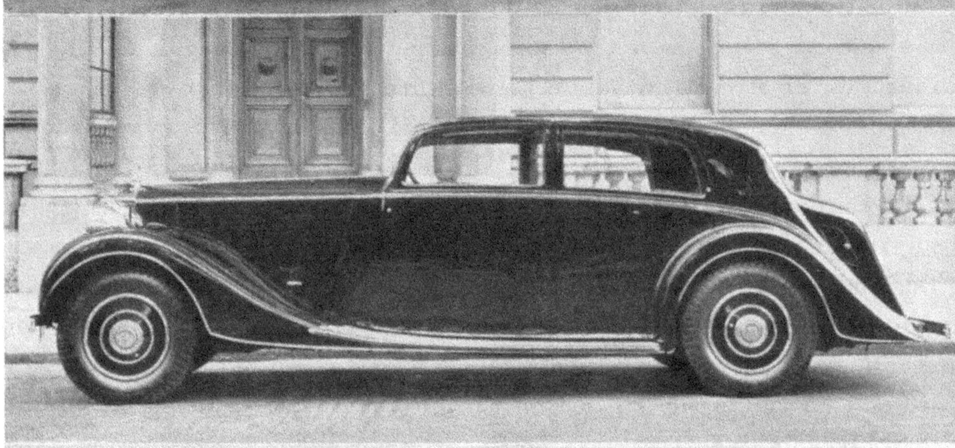

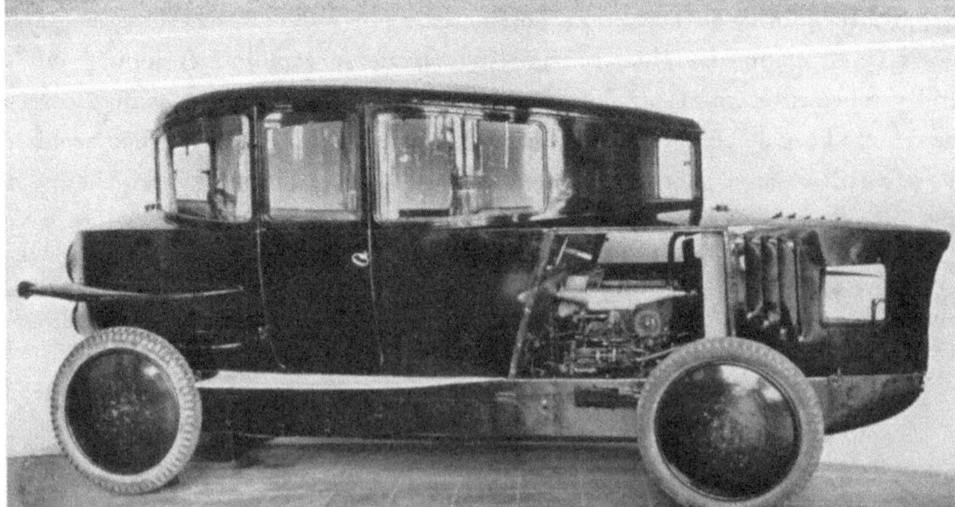

XXII. *Mercedes Benz 290 Cabriolet, long chassis, 1935–6. Price £995.*

Rolls Royce, Phantom III, Sports Saloon of 1936–7 designed by Jack Barclay, and made by H. J. Mulliner Ltd. Approximate price £2500–3000.

First streamlined car ever designed. Dr Rumpler's "Tropfenauto" of 1921 (engine behind, on the right). The original at the Deutsches Museum, Munich (by courtesy of the Deutsches Museum).

appears a somewhat funny relic. It will probably be several decades before motor-cars can be appreciated as objectively as pictures or pottery.

Generally speaking, the conception of streamlining is no doubt a legitimate symbol of our generation, a generation worshipping speed as a value in itself. It matters little that it is only a symbol. Though the public believes that streamlining has been introduced in order to increase speed, every one of the manufacturers whom I met agreed that the few miles per hour which might be gained by the slanting outlines of the car are not the true reason for the innovation. The reason, given time and again, was fashion. And since this fashion is so emphatically representative of a certain quality of this present age—a quality which does not seem to be of the kind that rapidly passes—it may be as well to use here the more serious and comprehensive term *style*.

Yet granted that streamlining is a true expression of the style of the twentieth century, the aesthetic value of the various forms which make a car appear streamlined varies greatly. Streamlined cars of totally convincing shape and proportion are in the minority. Most of the new models are designed to look somehow speedy and dashing and just left at that. How many unjustified chromium bands along the sides of car bodies, stuck on as a substitute for something more daring, do we see in the streets every day; how many freakish radiator fronts, concealing the real front of the radiator with bulging curves of some indistinct shape; how many cars the backs of which, owing to the massiveness of the cantilevered rear part of the body, are completely out of balance? Though intended to look light and fast, some of these cars remind one of the uncouth shapes of antediluvian animals.

Yet the most objectionable of these models are not to be seen in England. For even in so international a trade as the motor industry national differences of taste cannot be overlooked. England, averse to novelty and artistic adventure, has been much slower in taking up the new idea than America, France, Germany and Italy. Amongst the three largest mass-production firms in this country, the two really English ones have been extremely cautious, and it was Ford, the American firm, which pushed streamlining in cheap cars. So did Citroën in France and Opel in Germany, the largest manufacturers in those two countries. England's reluctance has not been confined to the inexpensive models. The same attitude seems to predominate in the upper as in the middle classes. To the French coachbuilder most of the work of famous British firms will appear slightly dull, to the English visitor the most elegant cars in the streets of Paris ridiculously flashy and showy.

I have not had many occasions in this book to praise the effect of British conservatism on industrial design. Here, I think, is one of them. The scepticism and steadiness that characterize English politics and English cultural life have saved the motor-cars of this country from most of the excesses of modernity to be seen on the Continent, although the same qualities may also—the inevitable opposite—have deprived England of some of the highest and rarest achievements.

Conservatism and sobriety are the most obvious general characteristics of English cars. Going more into detail, one cannot fail to observe great differences in style between the various price groups. To find examples of almost perfect balance and beauty, one has to look among cars selling at more than £600 or even more than £1000. Amongst medium-priced cars there are now several of very dignified and satisfactory appearance, less smart than corresponding Continental and American cars, but equally well or even better designed. I should include amongst this group products of Alvis (Firebird), Rover, SS., Riley, etc. They are characterized by a back which in outline consists of two curves, one for the back seats and one for the luggage locker, with a break in between, instead of the one long and daring curve which in foreign cars leads down from the roof to the back wheel mud-guards.

The situation is not quite as gratifying amongst the mass-produced inexpensive cars, although things have considerably improved within the last year. I do not know of any English model under £160 to the appearance of which I could give unstinted praise. The objection to be made against mass-production manufacturers is their reluctance to consider any innovations, a reluctance frankly admitted and accounted for by various reasons of varying validity. Second-hand cars, I was told, would be too rapidly devalued by a quick change. Another argument was that cheap cars must be as unobtrusive as possible, because a striking appearance, whether good or bad does not matter, always means publicity for the owner, and is therefore suitable only for expensive cars in which people want to be seen and admired. Furthermore the class buying inexpensive cars is supposed to be the mainstay of conservatism. Lastly, the production of a car at £115 or £120 leaves so little margin of profit that every endeavour must be made to eliminate anything which may increase costs. Therefore new parts of different shape must be avoided, and nothing is used which has not been tried out for some time and found satisfactory. The same applies to the assembling of the parts. Here again the design must demand no process which the worker has not got at his fingertips. For any change is bound to make itself felt in the speed of production. All this sounds convincing. And yet it still remains to be proved whether a frank avowal in favour of new and improved design might not induce a good many people to buy a cheap car who now, in order to have something pleasanter and more modern-looking, go on paying their instalments over a period somewhat longer.

I am afraid I have been applying to motor-cars severer standards than accord with the more lenient attitude adopted in other sections of this book. However, I think one is justified in demanding much, in demanding perfection, from an industry which has already attained so much. For directly one thinks of the atrocious carpets, or jewellery and silverwork, with which I had to deal, one cannot fail to see and admit that the motor industry is one of the best industries in the world as far as design is concerned. One sees only a few superfluous frills, no "arty" decoration either floral or cubistic (which for some inconceivable reason *has* been introduced into the design of perambulators), not much fuss, a clear and precise expression of function. The public knows what it wants, because it thinks more about motor-cars than about electric fittings or carpets, and the manufacturer, for his part, knows what he owes his job. Both parties to the industrial process, producer and consumer, here act reasonably and interestedly. But the lion's share of the praise is no doubt due to the manufacturer and not to the public. It is the attitude of the manufacturer (who, as I said, is at the same time the creative designer) which has determined the artistic standard of the motor industry. The manager of an automobile factory never seems to be the narrow-minded, badly-dressed provincial bourgeois whom one meets in so many other industries. Though living in a provincial town—it is known that Coventry houses the majority of English automobile firms—he has not been spoilt by the atmosphere. On the contrary, the spirit of so many modern factories has succeeded in changing the atmosphere of the town. Coventry is definitely industrial, but it has little of the dreariness of English nineteenth-century industrial centres such as Stoke. Here are men who have created an industry that is now one of the largest and most flourishing in the Midlands. Their adventurous and progressive spirit induced them ten, twenty, thirty years ago to back something new and untried. They know what a risk means and that it is only the taking of risks which makes life worth living. Why should they be afraid of a genuinely modern shape? They know, consciously or unconsciously, that they can rely on the public, and that what arouses their own enthusiasm will also prove the fulfilment of that public's secret wish. Speed is one of the leading motives of our age and of the whole of European civilization; the conquering of distance has been at the root of almost every new invention since the Middle Ages.

I do not think that these conclusions are too sweeping. If manufacturers who truly believe in their age will take the risk of pushing a new and unprecedented type of design, because they want to be proud of their work and would not feel happy in making money out of something that they could not approve of themselves, then success, artistic as well as commercial success, is irresistibly theirs.

ART SCHOOLS VISITED

My experience with manufacturers, retailers, designers, artists and architects has formed the material of the foregoing sections of this book. The next step in the investigation of the problems in question must be to give an account of the existing training establishments in this country which have made it their task (or one of their tasks) to educate for any form of activity connected with industrial art. Again, as in my inquiries into industrial production, my centre was Birmingham. Hence the Birmingham School of Arts and Crafts was examined more thoroughly than any other. I added evidence from the following schools around Birmingham: Kidderminster, Stourbridge, Brierley Hill, Wolverhampton, Walsall. Moreover it seemed to me necessary to complete the information thus received by seeing the Burslem Art School, the London Central School of Arts and Crafts, the Royal College of Art, the Manchester School of Art and College of Technology, the Leicester College of Art, the Shoreditch Technical

Institute, and the Wycombe Technical Institute. I also had the pleasure of discussing the matters in question with Mr Walter Keesey, Board of Education Inspector of Art Schools in the Midlands. To him as well as to the principals of the schools visited, above all to Mr H. H. Holden, I want to express my sincere gratitude.

From the outset it has to be stressed that my object was only one aspect of art education, namely the training of skilled workers and of designers for industry. All other aspects I had to leave out of account, even the all-important problem whether and how in the various schools the teaching of design and handicraft is linked up with that of painting and architecture. For this, as well as for the historical development of art schools in England and abroad, I must refer in anticipation to the last two chapters of a book on the history of Art Academies, which I hope will soon be ready for publication.

In more than one respect it was fortunate that my investigations had to proceed from Birmingham. For one thing, Birmingham has perhaps the most complete system of art school education in the country. The Central School of Arts and Crafts was founded in 1884 as the first Municipal Art School in England, and at the same time the first conceived under the influence of the Morris Movement. Under principals such as Catterson Smith that tradition was kept alive; and some years after the war the necessary step was taken from the almost exclusive consideration of handicraft to the satisfaction of industrial requirements. This was due to Mr J. B. Fletcher, one of the founders of the Design and Industries Association, who was called to Birmingham from Leicester. To-day (session 1935–6) the school has about 4900 students who attend courses in four departments, namely junior schools (259 students), the Central School of Arts and Crafts (1914 students), branch schools (2347 students; over 500 of them at Bournville), and the special School for Jewellers and Silversmiths in Vittoria Street (367 students).

There are two junior schools, providing three years' full-time courses for boys and girls between the ages of 12 and 16 years. Each year 72 boys and 24 girls who show particular skill in drawing and crafts are chosen from elementary schools. They receive 12 hours a week general education during the first, 9 during the second, and 6 during the third year, the rest of their time being devoted to a specialized trade training in drawing and either cabinet-making, or painting and decorating, or metalwork, or modelling, or printing, or fashion-drawing and needlework.

After three years they enter the trade for which they have been trained as skilled workers, and it has not proved difficult to find places for all of them. Only a selection of the very best pupils go on to the Central School. It is even more rarely that students from the branch schools move over to the Central School, the reason for this being that the courses at the ten branch schools are generally attended only by people who wish to do some drawing or craftwork as a spare-time occupation or in addition to their regular daily jobs. Almost all the classes in the branch schools are held in the evenings.

The School for Jewellers and Silversmiths in Vittoria Street was founded in 1890. It was first organized by the Jewellers' and Silversmiths' Association, but was later taken over by the Municipal Education Committee. It is all the more worthy of special attention, inasmuch as it is one of the very few "Fach-schulen" or monotechnics in the country, i.e. schools preparing for one industry exclusively, though it differs in many respects from the Central European monotechnics. Above all, the training is mostly done in evening classes on a voluntary basis. There are no full-time students, and few firms send their apprentices to the afternoon classes. The aim of the courses is to give workers in the trade an opportunity of completing their knowledge and skill, which in most factories are used only for routine work in one special process.

The Central School has the following courses: architecture and building construction; bookbinding; cabinet-making; commercial art, illustration and poster design; dressmaking and fashion drawing; embroidery and weaving; etching and woodcutting; gesso work and gilding; industrial and commercial design; leatherwork; lettering and illumination; life drawing and painting; metalwork, silversmithing and enamelling; painting and decorating; printing; stained glass; sculpture and modelling, teachers' training. The building is in the centre of the town and was opened in 1885. It is now no longer equal to requirements, and the erection of new premises is under consideration.

The prevailing taste in the various departments is naturally uneven. In some classes it is still conventional, in others, such as cabinet-making and printing, it is excellent. Equally variable is the relation of the courses to handicraft and industry. Bookbinding, leatherwork, and weaving are done with a view to handicraft only; silversmithing, though, or because, it is taught in the Vittoria Street School with an exclusive eye to the requirements of the trade, is in the Central School treated in a way more suitable for a future teacher or the amateur than for the industrial designer. In fact more than half the students of the class belong to those two categories. Collaboration with local trades, which is one of the foremost aims of the present principal, Mr H. H. Holden, has reached different degrees in different courses. The heads of the printing and the cabinet-making departments complained of the lack of interest on the part of firms. On the other hand, the School of Dress Design and Dressmaking has found ways to place most of its students as dress-designers, dressmakers, dress-advisers, cutters, etc. in Birmingham firms.

A municipal art school in a large English city is a very complex unit. In order to judge rightly the results obtained one has to be quite clear as to the various aims pursued. The Birmingham Central School trains artists, architects, commercial artists, craftsmen, skilled workers and designers. The first four categories as a rule attend full-time day courses. We are more concerned with the fifth and sixth categories, above all the sixth. What is done for the education of the industrial designer? So far as experimental craftwork goes—which is certainly necessary for the creative designer—most of the classes at the Central School would be of advantage; so far as the development of truly modern standards of taste goes, at least some of the classes. But as to the special requirements of design for machine production, not much has yet been provided. This is due partly to defects in equipment, partly to outlook, and partly to organization. For obtaining an insight into the processes of machine production, much more machinery ought to be available than the Birmingham School has hitherto been able to acquire. Only the jewellery school, and perhaps the printing department, have up-to-date plants. In some classes the attitude of the artist-craftsman still rules unchecked. In other courses, such as cabinet-making, the demand in the Midlands for special designers is too small to justify much stress on the designing side.

In order to provide education for the industrial designer without restricting the development of handicraft, a separate class for industrial design was established and is now directed by Mr A. E. Harvey, a distinguished free-lance designer to the metal trades. His students are drawn from the various craft classes, but since actual production of objects is excluded from their work, the scope of the class is gravely limited. The instruction given is confined to drawing on paper, and therefore remains more theoretical than is desirable. To make up for this, means should be found for sending students into factories, or for fruitful co-operation with the better equipped departments of the school. It might be very advantageous if a certain ascendancy of the design class over the craft classes could be introduced, so far as that is feasible; for there cannot be any doubt that the education of the designer will become more and more important amongst the aims of a live art school.

The spirit of the Birmingham School appears to be very satisfactory. And although many manufacturers and business men do not yet realize that work which might be useful to them is done at the school, things are evidently improving in that respect also. Teachers and principal seem equally open to suggestions, both from business men and from outside individuals.

The contrast between the Birmingham School and the Manchester Municipal School of Art seems to the visitor less a contrast of organization than of spirit. The school has about 1200 students who are trained in fine art and architecture as well as in crafts. There are junior courses, and there are trade classes, though less of these than in Birmingham.[1] No special class for industrial design exists. But the decisive fact about the school is that the emphasis appears to be laid on the "higher" arts and crafts, whereas—I could not help feeling this—the trade courses are regarded as a branch of less intrinsic artistic value. The classes are small, the dominant taste (except in a few courses such as printing and design for cotton printing) conventional. Of collaboration with industries I could trace very little. This is all the more regrettable because Manchester is, as everybody knows, the centre of an industry in constant need of designs and designers. The Lancashire cotton weaving and cotton printing trade certainly ought to

[1] 1935: junior courses, 58 students; general art course, including crafts: 183 full-time, 94 part-time afternoon, 496 evening students; trade courses: 65 part-time afternoon, 323 evening students (of these, 69 painting and decoration, 45 architecture, 39 cotton print design, 28 window display, 27 commercial art, 23 plastering, 22 dress trades, 21 wood and stone carving, 16 ticket writing—but woven designs 3 only).

have a "Fachschule" fully equipped and providing a comprehensive training for designers. Instead of that, there are a few poorly attended classes for textile designing in the Art School, and there is the Municipal College of Technology with its special Department of Textile Technology. Here scientific instruction of university standard is given in spinning and weaving, with the aid of a remarkable plant of looms, etc. The existing day course for designing is part of the ordinary degree students' work. Higher course students may specialize in design during their final year. This, however, was not done by a single student in the 1935–6 session. The evening course in design was attended by six second-year students, one third-year student, one fourth-year student of cotton manufacture and three of textile export, six fifth-year students of cotton manufacture and one of cotton industry. Hardly any of these students actually go into the trade as designers.

From this it appears that far too little is still done for a thorough education of textile designers.* A close co-operation between the Art School and the Technical School ought to be the first step. Such co-operation does not exist in Birmingham either (except for architectural students), but there it is less necessary, owing to the wide field of activity covered by the Art School. In Manchester it would be practicable without great changes of organization; and it is to be hoped that any personal difficulties, or difficulties with regard to outlook and principle, may be removed before it is too late.

I was not able to see any other Lancashire art schools nor any in the Yorkshire textile district, though I heard that some of them, Huddersfield for example and Macclesfield, are regarded as highly satisfactory in their relation to the local industries.

The same is true of Leicester, where the College of Art and the College of Technology are housed in the same vast and up-to-date building. Students taking the course in knitted fabrics are signed on at the Technical School but have to attend compulsory design courses at the Art School. Conversely, students of boot and shoe design belong to the Art School but have to go to boot and shoe manufacturing courses at the Technical School. Beyond the organization, the spirit and taste seem most gratifying, owing above all to Mr J. B. Fletcher who was Principal of the school from 1900 to 1920, i.e. during the time when the change from craft to industry became topical in all art schools. The Leicester College was one of the first English schools to realize the necessities of the new century, and it has succeeded in impressing its ideas upon more than one of the local trades. Leicester printing and lettering has for a long time been famous all over the country. It originated in the school, as did the project of developing a local cane-chair industry on the lines of that which flourished in Austria. Mr Fletcher induced the late H. H. Peach to take this up in 1907, and by 1914 the factory was giving work to about 200 people. H. H. Peach became one of the founders of the Design and Industries Association together with J. B. Fletcher.

As to the training of designers for the most important Leicester industry, namely hosiery and knitted fabrics, the school does very creditable work; and it has proved easy to place students at the end of their courses. Manufacturers are interested in the college, not only those in the textile trade, but also owners of printing works. The number of firms sending apprentices to afternoon classes and paying their fees is much higher than in Birmingham—a further proof that the untiring educational energy of an art school, continued over a sufficiently long period, can in the end convince employers. It may even—Leicester seems to point to this—change the appearance of a town.

The fourth Municipal Art School which I visited was the London Central School of Arts and Crafts. The problem here is different from that in any other city, because the size of London does not allow of the centralization found at Birmingham and Leicester. The London County Council has twenty-six schools giving instruction in arts and crafts. The Central School in Southampton Row corresponds to a certain extent to the Birmingham School, at least in so far as it gives mainly advanced courses. The other twenty-five schools either resemble the Birmingham branch schools, or are organized with a bias towards one special trade, or they work in conjunction with regional Polytechnics or technical schools. To give just a few instances, I may mention the North-Western Polytechnic with full-time junior classes in light engineering and dress-making, and evening classes in printing and cabinet-making; also the Chelsea School of Art, specializing in the graphic arts; the Fleet Street School of Photo Engraving; the Hammersmith School of Building and the Arts and Crafts; and above all the Shoreditch Technical Institute. This school, housed in very inadequate premises, which are going to be entirely rebuilt, is the nearest approach to a "Fachschule" of all the London training establishments. It consists of four departments: a training college for future teachers who want to specialize in handwork, a course preparing for the City and Guilds' Examinations in handicraft, a school for upholstery, embroidery, dressmaking and tailoring, and the school for the furniture and allied trades. In the furniture school, there were (session 1935–6) 121 boys between 13 and 16 years being trained for industry, and 393 senior students. The juniors as a rule study under County Council scholarships, and are at the end of their three years at the school easily absorbed by the trade. Of the seniors, 14 only attended the newly instituted day course for future designers, buyers or salesmen, all the rest being enrolled for evening classes. Nobody is admitted to these unless he is actually making a living by the trade. No firms, however, send their employees, nor are any fees paid by manufacturers. The course in cabinet-design was taken by 53 students, that in interior decoration by 30. The character of the cabinet-work which I saw was sound and attractive, whether done by juniors or seniors.

Following this mention of the Shoreditch Institute, a further parenthesis may be allowed. My visit to the Wycombe Technical Institute was also chiefly concerned with the special furniture courses held in their art department. At High Wycombe, Art School and Technical School are housed in the same old-fashioned building. The Art School is attended by about 50 full-time, and about 250 part-time students. A junior department prepares boys for both schools, and until the third year of the courses little distinction is made between their different needs. The senior students usually go into the trade, the most talented ones trying for jobs in drawing offices or for a scholarship to the Royal College of Art. However, the design course in 1935–6 was attended only by three or four students. Afternoon classes are frequented by apprentices who are allowed time off by their employers on one or two afternoons each week. Only rarely are the fees paid by the firm. Of the evening students—mostly workers, foremen, or draughtsmen—more than four-fifths take classes in cabinet-making, chair-making, furniture-drawing and upholstery. The equipment of the school is not bad, perhaps just sufficient for what the staff regard as their aim. A new building is planned. The taste seems to vary in different classes. Simple cabinet-work is quite sensible, whereas the schemes for interior decoration which I saw were mostly rather "tradey" and distinctly "modernistic".

After this digression on two schools of the monotechnic type, we must return to the London Central School, which is drawing its students from the other London schools as well as from art schools in England and abroad. It is housed in a building the scale of which appeared lavish when it was set up, but which is now so insufficient in many ways that the addition of another wing is under immediate consideration. The foundation of the Central School in 1896 was due to the influence of the Arts and Crafts Movement of which Lethaby, its first principal, was one of the leading brains.

The spirit of the school is extremely alive, and the artistic standard perhaps higher than in any other English school. The organization, however, seems strangely unbalanced, the result more of natural growth than of deliberate development. The total of students was 1801 in 1935–6. There are three-year junior courses for two trades only: book production and silversmithing. The teaching appears thorough, although I heard complaints from manufacturers of silverware that the instruction was more suited to the craftsman than to the worker in industry. I do not know how far this may be correct, but it is a fact that practically all the boys easily secure posts in their trades as soon as they have completed their third year. As to senior classes, only 10 per cent. of the students (190) attend full-time courses. Most of them want to go into some branch of commercial art (there is a considerable demand in London for artists engaged in this comparatively new profession) or become independent craftsmen. Of the part-time students, some are practising artists and craftsmen, but the majority work in trades. Very few of these are sent by their firms in the afternoons, owing partly to lack of interest on the employers' side, and partly to distance and cost of transport. In this respect, the size of London is a very serious obstacle to any comprehensive system of instruction in industrial art which is intended to embrace the whole city. The biggest furniture factories, for example, are in and near Tottenham. How can they send workers or draughtsmen to afternoon classes at Shoreditch or the Central School?

Of the 899 part-time students in artistic professions and industries 464 are artists, architects, designers, draughtsmen or teachers, 173 are in the printing, 156 in the silversmithing and jewellery, 60 in the cabinet-making trades. This shows that the school has not the same reputation in all its branches. As junior courses are confined to printing and silversmithing, only the trade classes for those two industries are sufficiently equipped and sufficiently closely connected with the employers. The textile class has no power-loom, the glass class no blowing or engraving apparatus.

It would be necessary to develop these and other sections, if it were decided to make the school the centre of London's industrial art education. Whether such a scheme of centralization is desirable is a question to be discussed in Part II. At present it is sufficient to say that if a real Central College is to be established no other institution would be equally suitable, for the spirit of the London Central School would guarantee a modern and sound training in any course that might be added or expanded.

Apart from these four big Municipal Art Schools with which I have dealt in the foregoing pages, I tried also to get some insight into the relations existing between local industries and art schools in five smaller towns in the Midlands. What do the Stoke Schools do for pottery, the Kidderminster School for carpets, the Walsall School for leather, the Stourbridge and Brierley Hill Schools for glass?

Stoke-on-Trent has four art schools, of which three run evening classes only. I was shown round the fourth, the Burslem School. This has a full-time junior department, courses for teachers, evening classes for workers in the trade, and also a design course. Almost all the members of the staff are engaged in the local industry. The school is by no means a "Fachschule", though its Principal, Mr Gordon Forsyth, is endeavouring in every way to develop the pottery courses. He is a trained potter himself who has worked for years in the trade. The equipment which he has at his disposal is far from adequate; firing can be done only on a very small scale. But in spite of that, the school has succeeded in securing the interest of the best manufacturers. Many firms send apprentice boys and girls to the classes and pay their fees, and the high standard maintained by the Principal and some of the teachers has remarkably increased the artistic value of the production in various works. Mr Forsyth's own sympathies in pottery decoration seem to be rather with the development of modern hand-painting than with mass-production methods. It is noteworthy that hand-paintresses trained by him have been able to hold their own against the progress of unskilled labour in pottery making, and that—as I mentioned before—several unscrupulous manufacturers feel annoyed by this human competition with their mechanical methods. Among the most progressive and conscientious industrialists, however, he is fully recognized. One of the best of the younger potters, Susie Cooper, was his pupil. While she and some manufacturers influenced by his style, and supplied with former students of his, have developed a very attractive style of decoration, the taste inculcated by some of the teachers in the Stoke schools appears, to judge from what I saw at exhibitions, rather reactionary and, what is worse, "tradey". Thus the excellence of the Burslem School and its influence on industry appears to be to a large extent the work of one personality rather than of any special system of organization.

And this brings me to the main reason for the weakness of the Kidderminster Art School in dealing with the carpet industry. I could not find much fault with its organization. It is not a "Fachschule", as perhaps it ought to be, but it has a special course in carpet design. This course is well attended by employees of local firms, the manufacturers being quite willing to give their youngsters an afternoon off and pay their fees. The equipment, however, is preposterously inadequate; no apparatus for hand-weaving or hand-knitting or for Axminster and Wilton processes, is at the students' disposal. But it is doubtful whether much could be gained by any improvements of plant, as long as the taste prevailing in the class for carpet designers is in no way above the standards of the Kidderminster trade.

The situation is in certain ways even worse at Walsall, though the Principal of the Art School is modern-minded enough to desire a close co-operation with the local fancy-leather trade. He complains of the almost complete lack of support from manufacturers. Some of them do not even allow him to see their works, for fear, I suppose, of being robbed by him and his school of their precious designs. Another argument used by producers against an all-round training in leatherwork, such as the school wants to give and could easily give, is that it would encourage students at the end of their course to settle down and open new firms for which, as I said before, very little capital is needed. The organization of the school seems satisfactory. It was one of the first in England to set up a full-time junior course, and has about 300 apprentices in its evening classes. The prevailing spirit also seems to be alive. But what is the good of that, if the most important local industry does not realize in the least what benefit it could derive from the work of the school? It is not at all unlikely that such considerable improvements of design as the school could gradually bring about might help Walsall in resisting the growing ascendancy of London. And here, for once, improved taste would in all probability mean immediately improved sales.

The conditions in the centre of the English glass industry are similar in several respects. The Birmingham Central School teaches only stained glass designing and making; and the two local art schools at Stourbridge and Brierley Hill struggle against the lack of help and understanding on the side of the producers and against entirely insufficient equipment. Hardly any firms send their apprentices; the plant consists of engraving lathes and nothing else. No firing, no blowing, no cutting can be done in the schools. The glass classes are attended at Brierley Hill by about ten to twelve students, at Stourbridge by three in the engraving and four in the designing classes. The impression on visiting the two schools is all the more depressing, because, at least in Stourbridge, it is evident that an active and enterprising spirit prevails.

It is not easy to sum up the results of my visits to those eleven Art Schools. The most characteristic feature of English as compared with typical Central European Art Schools is that their activities are so much wider and more varied.[1] A completely developed English art school would run a full-time junior department, preparing boys and girls of 12 to 16 years for certain skilled trades, afternoon courses and evening courses for skilled workers, craftsmen and some designers, craft classes for amateurs and future independent artist-craftsmen, art classes for future artists, and courses for teachers. That means that one institution in this country fulfils the same purpose as four or five different schools in Germany—namely Fortbildungsschulen, Gewerbeschulen or Fachschulen, Kunstgewerbeschulen and Kunstakademien.

[1] With this and the following remarks cp. A. Abbott, *Trade Schools on the Continent*, Board of Education Educational Pamphlet No. 91, London, 1932; E. M. O'R. Dickey and W. M. Keesey, *Industry and Art Education on the Continent*, Board of Education, Educational Pamphlet No. 102, London, 1934; and also F. A. Taylor, "Art Training for Industry on the Continent", *Journal of the Royal Society of Arts*, vol. lxxxiv, pp. 64ff., 1935.

It would be wrong to give preference to either of the two systems. The tendency in Germany since 1900 has been to yoke together the training in Fine Arts, Crafts, Industrial Design and Industrial Production. The "Fachschule" combines the standards of first-rate craft schools with the commercial and technical training given by a technical or trade school; and the schools of arts and crafts tend more and more to incorporate fine art. In fact the two most remarkable events in the history of German art schools since the war, the foundation of the Bauhaus and the union between the Berlin Academy and the Berlin School of Arts and Crafts, were based on that desire for a closer co-operation between architecture, art, craft and industry.[1] At the Bauhaus, Walter Gropius had made architecture and the building crafts the central feature of the training. Handicraft was regarded mainly as "being responsible for the experimental work for industrial production". In the curriculum from the preliminary course onward equal importance was assigned to design and to handwork. The object was for the workshops to be "laboratories in which utensils typical of the modern age and ripe for series production were carefully evolved in the model". Students had, during the early years of the Bauhaus, to pass the municipal journeyman's examination before they were admitted to the highest classes and to actual participation in the Bauhaus's building tasks. A similar synthesis existed at the Berlin Schools, where no artist was admitted unless he was willing to undergo a thorough craft training first.

In England the dominant tendency to-day is exactly the opposite. Local art schools try to develop into monotechnics; separate continuation schools are recommended. But the most modern and live education for future painters and sculptors is given in some of the schools of arts and crafts, such as the London Central School, or, to mention also a small school, the Art School which is part of Goldsmiths' College at New Cross. In such circumstances it would be narrow-minded to recommend a change of direction. More detailed suggestions as to the course which might be followed in this country will be given in Part II.

So far as co-operation with industry is concerned, the results achieved have

[1] Cp. N. Pevsner, "Post War Tendencies in German Art Schools", *Journal of the Royal Society of Arts*, vol. lxxxv, p. 247 ff., 1936; W. Gropius, *The New Architecture and the Bauhaus*, London, Faber and Faber, 1935; and W. Gropius, in *The Year Book of Education*, pp. 493 ff., 1936.

proved very different in different towns and trades. The training of skilled workers for silversmithing and jewellery in Birmingham, the training of skilled workers and creative designers for pottery in Stoke, or for printing in Leicester and London, the training of designers and workers for clothing firms in Birmingham are excellent, while for the glass and leather industries arrangements are still totally inadequate. To understand the situation one must always remember that the English art school is at the same time the training place for workers and for artist designers. Now it must be said that, generally speaking, a great deal has been done hitherto for the all-round training of the worker but only a little for the education of the creative designer. Evening courses for skilled workers are usually well organized and well carried out, corresponding afternoon courses are all too often impossible, because manufacturers are not sufficiently interested. Their lack of interest is in most cases due to ignorance of the new tendencies in art schools, which are still considered "crafty" and unpractical. Also a great number of manufacturers dislike art schools, because they want "hands", and have no use for human beings with an all-round training. Such a continued reluctance of producers to give a chance of furtherance to their young employees seems to me one of the cases where government interference is necessary; but that point again has to be left to a later page.

As to special provision for industrial designers in schools, things are certainly still in their infancy. Very often, indeed, no clear differentiation is yet made between the needs of the future independent craftsman and those of the future industrial designer who wants instruction in handicraft for experimental purposes only. The idea of a separate class for Industrial Design such as has been introduced in Birmingham appears very promising, provided that it can be given sufficient authority amongst the existing crafts classes to secure an appreciative execution of the designs produced. However, all progress in courses of that kind is gravely hampered by the small demand for well-trained designers in most trades, and consequently by the poor provision, in most schools, of machinery for modern volume-production. Here again support from manufacturers, trade organizations, municipal and State authorities is badly needed. Now and then, co-operation with existing technical schools could make up for the lack of certain machines. More often, I am afraid, only changes in personnel and the appointment of men who at the same time are of a high cultural status, have full experience in industry, and possess good and modern standards of taste would induce manufacturers to take more interest in the schools. The fact that men are more important than organization is valid here as well as in those industries which have been discussed.

As to the new men, the situation seems hopeful. The policy which the Board of Education inaugurated by the issuing of Circular 1432 to all art schools at the end of 1932 has proved successful in several recent appointments of inspectors, principals, and teachers, and the excessively urgent reform of the Royal College of Art seems to be approaching at last. The Royal College, which is the only government art school in England, was founded in 1837 in order to extend "a knowledge of the arts and of the principles of design among the people", using "not theoretical instruction only, but also the direct practical application of the arts to manufactures".[1] In spite of this precisely defined programme, the College during the nineteenth century became more and more a school for the training of drawing masters. Relation to crafts and design were forgotten, and even the Morris Movement could not for several decades change this attitude, which is so typical of the century of unchallenged Liberalism. It was not until 1899, when Walter Crane became Principal for a short time, that new regulations were introduced which led to the establishment of a Design Department. Lethaby, first Principal of the London Central School, was made Professor of Design and remained from 1900 till 1918. But the new impetus cannot have carried very far, for in 1911 a Government Committee pleaded for the total dissolution of the College, and in recent years the Design and Industries Association, as well as a specially appointed Government Committee on Advanced Art Education, still had good reasons for proposing a complete reorganization in order to restore to the College its *raison d'être*.[2] How far the changes now planned will go one cannot tell; at any rate the appointment of Mr P. H. Jowett, Principal of the London Central School until 1935, as Principal of the Royal College makes it more than likely that changes there will be.

[1] Cp. *Report of the Departmental Committee on the Royal College of Art*, pp. 27 ff., London, 1911.

[2] Cf. *Design for To-Day*, vol. ii, pp. 74ff., 1934, and Report of the Committee on Advanced Art Education, H.M. Stationery Office, 1936. On this *vide* *.

In these circumstances, any detailed criticism of the serious deficiencies of the college would be unfair and unnecessary. A very short account of the structure of the Royal College, as it was when I visited it, is all that is required. There are five separate departments in existence, the Schools of Architecture, Design and Crafts, Painting and Drawing, Engraving, and Sculpture. The School of Design has courses in pottery, metalwork, stained glass, cotton printing, weaving, embroidery, wood-carving, decorating, lettering and illumination, and book-binding. The equipment appeared scanty and hardly suitable for instruction in industrial methods. No clear division between the teaching of handicraft and design was made; but the spirit of the craft classes was cheerful and adventurous. The number of students is considerable. Most of them come from provincial or London art schools, the majority with scholarships. It is worth pointing out that, although up to the present the emphasis has undeniably been placed on the education of painters and teachers, the Design School is by far the largest department of the college. The following are the numbers of the students who entered the school in the seven scholastic years between 1930 and 1937:

	Design	Painting	Sculpture	Engraving	Architecture
1930–1	79	41	5	2	–
1931–2	68	48	8	10	–
1932–3	60	38	5	6	–
1933–4	75	28	2	6	1
1934–5	73	30	6	6	1
1935–6	67	31	6	9	1
1936–7	47	24	9	4	–
1930–7	469	240	41	43	3

These figures clearly show the necessity for an immediate reorganization of the Royal College on the basis of industrial needs. Practical difficulties certainly ought not to be underrated, but several improvements are feasible and appear almost within reach. Part II will contain more about this.

Before passing on to conclusions, it is necessary to compare my results with the work done by other institutions and individuals on similar lines or with similar aims. I shall deal with governmental committees and private societies first, with individual enterprise and publications afterwards.

THE BRITISH INSTITUTE OF INDUSTRIAL ART

The British Government realized as early as 1914 the necessity for some organized effort in order to improve industrial art. Immediately before the war the Board of Trade and the Board of Education decided to found an institute of a semi-official character, whose main activity was to be the arrangement of exhibitions. The war prevented the realization of these plans, but the work was taken up again in 1919, and in February 1920 the British Institute of Industrial Art was incorporated. Sir Hubert Llewellyn Smith was designated as chairman. A Treasury grant enabled the Institute to start in a conspicuous way almost at once. In May 1920 a Gallery for exhibitions of industrial art and handicraft was opened in Knightsbridge. Its principal aim was to find a market for the work of designers and craftsmen worth official support. During 1920 three exhibitions were held, from which the nucleus of a future permanent collection was purchased and for many years shown to the public in the North Court of the Victoria and Albert Museum. Now it has been removed to the Bethnal Green Museum, where it is seen by fewer visitors than it deserves. Owing to the first post-war slump, which became so menacing in 1921, the Government grant was not renewed, and the Institute, started with so much energy and impetus, became dependent upon private donations and subscriptions, though it maintained its semi-official status. Considering these economic difficulties, its achievements are worthy of the highest praise. The Gallery in Knightsbridge had to be given up, but various other exhibitions were arranged in the Victoria and Albert Museum and in provincial towns. Amongst the most remarkable of these was the large Travelling Exhibition shown in Bradford, Birmingham, Huddersfield and Leicester in 1922, and the exhibition in the North Court of the Victoria and Albert Museum which was held in 1929 under the remarkable title: Industrial Art for the Slender Purse. It was perhaps the first public recognition of the fact that the most urgent need is to improve the artistic quality of inexpensive objects. The other main activity of the Institute was research. Its pamphlets on *Industrial Art and the British Manufacturer*, on *The Training of the Middleman*, on *Public Departments and Industrial Art*, and on *The Training of the Silversmith*, were very well done and still remain the foundation of some of the most recent investigations. Since the Government had withdrawn from the struggle for the improvement of industrial art in 1921, private institutions became indispensable. The Design and Industries Association, with which I shall deal presently, was founded as early as 1915, the Industrial Art Committee of the Federation of British Industries in 1921. The Board of Trade and the Board of Education remained inactive until 1932, when the Board of Trade appointed the Gorell Committee, and the Board of Education sent out its Circular 1432 to art schools which was to be the beginning of a "New Deal" in art education.

THE INDUSTRIAL ART COMMITTEE OF THE FEDERATION OF BRITISH INDUSTRIES

The year 1921, when the Federation of British Industries decided to set up an Industrial Art Committee, was the same year in which the British Institute of Industrial Art had to restrict its activities. The Committee, which from its foundation has been under the chairmanship of Mr Charles Tennyson, has done very useful work in many respects. The following are some of its activities. In 1930 the Committee issued a memorandum on State-Aided Art Education which was partly responsible for the proposals contained in the Board of Education circular to art schools which I mentioned just now. There followed the organization of an Employment Bureau for industrial artists which has been used by some hundreds of firms. Jobs are found for 50 to 80 students a year, half of whom are commercial artists. Free-lance designers are also brought into touch with manufacturers. The bureau is provided with a register of suitable free-lance artists and a card-index of firms which might be interested in designers.

The Committee also succeeded in arranging refresher courses for industrial designers at the Royal College of Art, in promoting regular visits of art students to factories, and in the distribution of instructional notes on several industries to art schools. In 1935 a detailed memorandum was worked out dealing with industrial art training for the management staff of firms—both on a full-time and a part-time basis—and with a similar training for foremen, travellers and salesmen. Courses were suggested for entrants into a trade as well as for those already employed, the entire scheme being linked up with the reorganization of the Royal College of Art and the institution of regional colleges as set forth by the memorandum of 1930. Another memorandum of the Committee deals with the restoration and preservation of ancient buildings in England, calling attention to the importance of this question for any revival of handicraft and reduction of rural unemployment. A very sound report was also brought out on the Burlington House Exhibition of 1935.

THE GORELL COMMITTEE

Our survey of trades has brought to light the fact that decisive improvements in most industries date back no further than 1930. What the British Institute, the Federation of British Industries, the Design and Industries Association and private individuals had done before that time may be of the greatest historical value as pioneer work, but these efforts had remained isolated and sporadic. For the first twelve years after the war one can hardly speak of a live Modern Movement in English architecture and industrial art. But by 1930 animated discussions had begun in many quarters, and various activities were started under the auspices of numerous institutions—whose work, by the way, frequently overlapped and yet remained carefully independent, as is the habit in England. When things had gone so far, the British Government decided to return to the fray.

In 1931 the Board of Trade appointed a Committee on Art and Industry, under the chairmanship of Lord Gorell. Among its members were personalities such as Roger Fry, A. E. Gray, C. H. St John Hornby, Sir Eric Maclagan, H. Robertson, Sir Hubert Llewellyn Smith, H. Trethowan, Professor E. W. Tristram and C. Williams Ellis. This first official body did not concern itself with the whole intricate complex of problems. Some palpable results had to be brought before the nation as soon as possible. Hence, with wise self-limitation, the Committee confined its activity to examining the one question of Industrial Art Exhibitions, though this could naturally not be done without some side-glances at related questions.

The report of the Committee was published in 1932. It begins with a succinct survey of the history of collaboration between industry, commerce and art from the foundation of the Royal Society of Arts in 1754 until the war. It then sketches out some requirements for art education, rightly stressing the point that nothing conclusive can be done unless the citizen is brought up to be far more art-conscious and design-conscious than hitherto. Passing on to the actual object of the report, preliminary investigations into the conditions of art in various industries are recommended. In conclusion, the Committee, in virtue of the foregoing discussion, strongly supports the idea of a permanent exhibition of industrial art in London and of travelling exhibitions in the provinces. It is not necessary here to go into detail as to the ways recommended by the Committee for carrying this out.

The most important fact about the Gorell Report is that it was the first official document to emphasize the vital importance of improvements in British industrial art and to confirm the urgent necessity of immediate action. It is a carefully thought out and carefully written document, in spite of some deficiencies in wording which, to my mind, are overstressed in Mr Herbert Read's book on *Art in Industry*.

THE PICK COUNCIL

Following the suggestions which were made by the Gorell Committee, the Board of Trade appointed a Council for Art and Industry with Mr Frank Pick as chairman. Among its members there are industrialists, merchants, architects, artists and art critics. The Council started its activity at the beginning of 1934. During the first year such a comprehensive programme was worked out that a long time will be required to complete the task which the Council has set itself. The various objectives are mentioned in the Second General Report of the Council which was published early last year.[1] They are enumerated under three headings: education of the consumer in the appreciation of designs, training in design for those engaged in dealing with it throughout industry and commerce, encouragement of good design especially in relation to manufacture. Besides the English Council there is a special Scottish Council at work. The work is not limited to meetings, lectures and discussions; in contrast to so many councils and committees, this one has almost immediately begun to act with great energy.

[1] Cf. *The Board of Trade Journal*, 23 April 1936.

For the education of the consumer, special research was carried out in 1934 and 1935. The results were published as an illustrated pamphlet of 40 pages, dealing in a very concentrated manner with art education in elementary and secondary schools. Further, the Council has, a few months ago, held under the auspices of the London County Council an exhibition of educational material available for elementary schools, and—what is more—has been invited by the same body to plan and equip a new elementary school to be erected at Peckham. It is expected that the school will be completed in 1938.

With regard to the existing conditions of training for industrial personnel, evidence was collected from twelve trades.* A first full report on the silverware and jewellery trades has been announced for March or April. Special committees to co-operate with local education authorities in the replanning of industrial art education have been set up at Manchester for the cotton and rayon industries* and at Stoke for the ceramics industry.

Like the Gorell Committee, the Council had from the outset emphasized the importance of exhibitions and public collections for the education of the designer, the consumer, and the manufacturer. Here, again, almost immediately, action was taken. A small exhibition of silverware was held at the Victoria and Albert Museum in 1934 showing objects of the past side by side with modern examples. A pottery exhibition, seen by about 20,000 visitors, followed in 1935. In the summer of 1936 an instructive collection of domestic metalwork was on show. This year the Council will be responsible for the British pavilion at the International Paris Exhibition, and, so far as can be gathered from rumours and unofficial remarks, there seems to be some reason for prophesying a success. Plans are also under consideration for the arrangement of travelling exhibitions in the provinces, for a private show of exemplary foreign products of one industry, for the establishment of a National Register of Industrial Designers,* and—probably more important than anything else—for an exhibition of the complete furniture and other equipment of a four-roomed dwelling at an inclusive cost within the means of the working classes (£40–50, I understand).*

It is evident that any one of these tasks coincides with one or more of the problems discussed in this book. Moreover, as the means at the disposal of the Council are so much more extensive than those available for my research, as the evidence on which they will base their reports is bound to be so much more comprehensive than mine, and as their chances for translating into practice what they advocate are so good, while mine were almost nil—it must appear questionable whether this book can claim any *raison d'être* at all. I should not have started on it, however, had I not been of the opinion that there is at the present moment a certain scope for private research. For one thing, the reports of the Council will of necessity take a long time to prepare, whereas an immediate discussion of the whole problem as it is, based on as much evidence as is available to-day, seems extremely urgent. Also a book like this can be much more outspoken than a government report, and the information to be obtained in a private conversation may often be of a more intimate character than that which one would give to an official committee. However, there cannot be any doubt that in the end the chapters and sections of my book will be superseded by the reports and the other activities of the Pick Council, and I shall be the last to regret this.

THE DESIGN AND INDUSTRIES ASSOCIATION

More than fifteen years before the establishment of the Gorell Committee, and about five years before that of the British Institute of Industrial Art and the Art Committee of the Federation of British Industries, a private association had been founded for the promotion of art in English industries—the Design and Industries Association—to whose persevering propaganda probably more progress is due than to the activity of any other institution or individual. In spite of this, the association is still far from holding with industrialists, distributors and the public the position which it deserves, and which the German "Werkbund" had attained shortly before and immediately after the war. Yet the effect of its incessant work in creating a more favourable atmosphere can hardly be overrated.

The D.I.A. was founded in 1915. Its avowed model was the "Deutscher Werkbund", created in 1907. It is a remarkable instance of English sobriety and soundness that just one year after the outbreak of the war this dependence on a German model was frankly admitted. I wonder whether in any Continental country anybody at that time would have ventured to say in public: "When an enemy has a noble lesson to teach it can only be learned from him nobly."[1] On the first Council there were men like Sir James Morton, H. H. Peach, Frank Pick and Sir Frank Warner from industry and commerce, C. G. Brewer, J. B. Fletcher, Ambrose Heal, C. H. St John Hornby, F. E. Jackson, W. R. Lethaby and H. Stabler from art and architecture. The first pamphlet published by the D.I.A. in 1915 already contained the famous thesis, so much discussed since then: "The first necessity of sound design is fitness for use." One year later another equally significant tenet was formulated. One of the paramount aims of the Association should be "to harmonize right design and manufacturing efficiency, accepting the machine in its proper place, as a device to be guided and controlled, not merely boycotted". This was evidently meant as a challenge to the anti-machine attitude of Morris and his followers; and as such it still sounded revolutionary in England at the time when it was put forward.

After this brilliant start, the Association developed but slowly. The "Deutscher Werkbund" seven years after its foundation was strong enough to organize the most important exhibition of industrial art held in this century, the Cologne Exhibition of 1914, a token of the victory of the Modern Movement in Germany. The D.I.A. remained an obscure, though by no means ineffective, body for more than ten years. Membership gradually grew from 292 in 1916 to 602 in 1928. A small journal began to appear in 1917, but was not a success. From 1922 onwards there was published a series of excellently written and well-illustrated Year-Books plainly reminiscent of the pre-war Year-Books of the "Werkbund". In 1927 the very satisfactory English Section of the International Exhibition of Industrial Art, held at Leipzig, was organized by the D.I.A. under the direction of H. H. Peach.

Between then and now, many activities have been undertaken. Branches and local groups were founded in Birmingham, Bristol, Manchester and Northampton. Lectures were held; pamphlets and books, such as the extremely bracing and useful Cautionary Guides to several English towns, or the manifesto on the Village Pump, were published. The year 1932 saw the first adequately produced periodical of the Association. It was called *Design in Industry*. After a year it was replaced by a new periodical, *Design for To-Day*, which was never the society's

[1] A. Clutton Brock, *A Modern Creed of Work*, p. 8.

own paper but contained its official notes and remarks and was recommended and supplied by the society to its members. At the beginning of 1936 another magazine was started, *Trend in Design of Everyday Things*, which, although it was equally interesting in text and layout, had to be discontinued after only two issues had appeared—a depressing sign of the lack of public interest in the aims of the association.

The most remarkable action of the D.I.A. in recent years is connected with the Dorland Hall Exhibition of 1933. Though none of the official work was done through the office of the Association, and though its name is not to be found anywhere in the catalogue, it must be given the credit for the initiative as well as a great deal of the necessary spadework. In fact, the majority of the Council were members, some even in leading positions. The Exhibition, the best of its kind hitherto held in England, was an outcome of the suggestions made in the Gorell Report. A few more words will be said about it later. Part of the Exhibition was afterwards shown at Manchester, Halifax, Leeds, Edinburgh, Bristol and Stoke-on-Trent, admittedly under the auspices of the D.I.A. The Birmingham branch arranged an independent exhibition of Midland Industrial Art in the summer of 1934, mentioned before in connection with my questionnaire. The standard was by no means equal to that of the Dorland Hall Exhibition. Since then several exhibitions have been held based on another principle. The D.I.A. induced big shops or department stores to allow exhibitions on their premises, consisting entirely of articles that these shops have in stock and selected by the D.I.A. Such exhibitions, the educational value of which appears beyond doubt, have been held at Bowman's and Harrod's in London, Kendal Milne's at Manchester, Dunn's at Bromley and Rowntree's at Scarborough.

Another means of propaganda has been the exhibition of model houses completely furnished by the D.I.A. at a given cost. Such exhibitions took place at Manchester in 1933, at Welwyn in 1933, and at Birmingham in 1934. Here, of course, as well as in the exhibitions in stores, the aim was to impress upon the public that beauty in everyday objects is attainable at reasonable prices, if one seriously tries to find it. At Welwyn and Manchester the complete contents of the house did not cost more than £200. Of considerable merit also were the various exhibitions of photographs of modern architecture and industrial art, arranged by the D.I.A. and shown in London and the Provinces, not only in museums, but also in schools, and once—at the time of the B.B.C. talks on Design—in the Charing Cross Underground Station.

This survey of the activity of the D.I.A. will, I hope, have brought out what valuable work it has done, but also that the work is still far from complete. Membership in 1935 was only 819, of whom 110 were manufacturers, 101 designers, 98 architects, 93 teachers, etc., 57 painters, 55 retailers, 14 craftsmen. A greater number of seriously interested members would mean more money, more money would mean a more intense and widespread activity, and more activity in publications and exhibitions would mean a higher standing of the Association in popular and official recognition, which in its turn would attract more members and increase again the commercial and educational value of its propaganda.

THE SOCIETY OF INDUSTRIAL ARTISTS

The Society of Industrial Artists was founded in 1930 "to secure a recognized status for artists working for industry and commerce, and to provide a controlling authority; to advance and protect their interests". According to this programme, the Society might have developed into something like a Trade Union. Some of its activities in fact tend in that direction. A Register of Industrial Artists is planned (this was remarked as one of the aims of the Pick Council and the Industrial Art Committee of the Federation of British Industries). A code of basic rules for professional conduct is to be set up, advisory service to members is provided, and debts are collected. But besides that, the Society from the beginning considered its task to be the promotion of improvements in commercial and industrial art education, the organization of exhibitions and the fostering of a high standard of industrial design.

Consequently, an entrance jury was established which, however, was given up two years ago. The Society grew slowly. It had no more than 250 members in 1936. But since names like Chermayeff, Curwen, Dobson, Milner Gray, Jowett, Paul Nash and Sutherland are to be found amongst the members of the Council, its seriousness is evident. Most of the Society's activities are akin to those of the D.I.A., with whom a scheme of collaboration has been worked out, in so far as it consists of preparing the ground and creating an atmosphere favourable to improvements. The Society participated in the Dorland Hall Exhibition of 1933, has acted as a witness to the Pick Council, and is now beginning to develop local branches, of which the one at Stoke-on-Trent seems to work very effectively.

It is not quite clear what reason there was in 1932 for the formation of an Association of Artists in Commerce with a programme surprisingly similar to that of the Society of Industrial Artists. The success of the new Association cannot yet be judged.

THE ROYAL SOCIETY OF ARTS

The Royal Society of Arts was founded in 1754. Its aim was to be, in accordance with the ideas of the Age of Reason and of Mercantilism, "the Encouragement of Arts, Manufactures and Commerce in Great Britain". The Society has recently been blamed for not having done more to achieve a close co-operation between art, industry and commerce. But it is a misunderstanding to assume that the original programme intended any such co-operation.[1] In fact the Society was quite in harmony with the ideas of its time in keeping its various activities separate from each other. Scientific research was started to help industry and commerce. Art and industrial art could—this was the conviction of the leading theorists of the neo-Classic movement—be fostered only by promoting academic figure drawing, painting and modelling. Only thus could any reaction on the lower strata of commerce be conceived. Space does not allow me here to go more closely into the historical aspect of this idea.[2] Ideas had changed when in 1849 the society—mainly owing to Sir Henry Cole—took the lead in helping Prince Albert to prepare the first International Exhibition. The 1851 exhibition was in the end not in any explicit way connected with the society, but in spite of that its share in the preparatory work remains noteworthy, although the grafting of decoration on to industrial products as it characterizes most of the

[1] Cf. H. T. Wood, *A History of the Royal Society of Arts*, pp. 152ff., London, 1913.

[2] Once more I have to refer to my book on the history of art academies which I hope will soon be ready for publication.

exhibits of 1851 is by no means what we should to-day call a satisfactory co-operation of "Arts, Manufacture and Commerce". It is hardly necessary to add that one should not put the blame for this on the society, which was once more acting in harmony with the prevailing ideas of the time.

A decisive change came with the Morris Movement. Since Morris had preached the importance of art in everyday life, a new outlook on the problem of industrial art became possible. Now, the Royal Society might have been re-modelled according to the new standards. But this chance was missed, and the Arts and Crafts Movement spread independently of the Royal Society. A small number of annual prizes for students of design in art schools had been established in 1846, and these remained the Society's only contribution. As to the step from handicraft to machine-art, this was taken first by a few progressive individuals, and then by the circle of artists and manufacturers united in the Design and Industries Association, while the Royal Society still held aloof. In 1924, at last, a new departure towards the promotion of industrial design was made. The initiative was principally due to Sir Frank Warner, who has been mentioned before as the head of one of England's most distinguished textile firms. Funds were raised, and from 1924 until 1933 competitions for textile, furniture, book, pottery and glass design were held. They were open to all British subjects, not to art students only. In 1933 the competitions were given up, and the Society concentrated on the preparation of the exhibition at Burlington House which originated as a joint enterprise of the Royal Academy and the Royal Society. A separate section must be devoted to this exhibition.

THE ROYAL ACADEMY AND THE EXHIBITION OF BRITISH ART IN INDUSTRY AT BURLINGTON HOUSE

So much adverse, and in some cases malicious, criticism has been published of the Exhibition of British Art in Industry held in 1935 at Burlington House, that the primary duty of the historian is to report those facts which one has to know in order to judge of the organization and the results of the exhibition. First of all, it was a revolutionary act on the part of the Royal Academy to open its rooms to an exhibition of industrial art, and it must be recognized as such. It must also be recognized that the Royal Society of Arts by supporting an exhibition of this kind has clearly shown its earnest desire to help art where help is most urgently needed to-day. This should not be forgotten, whatever one may feel bound to criticize in the exhibition itself.

The following were given as its chief objectives: "To impress upon the British as well as the foreign public both the importance of beauty in the articles they purchase, and the fact that British Manufacturers in co-operation with British Artists, are capable of supplying such articles in all branches of industry." The other aim was to enable manufacturers to study to what extent British artists can supply "original, attractive and technically suitable designs for the production of articles by mechanical means". The general conditions stated that all exhibits must be of original, contemporary British design, created by designers of British nationality and working in Great Britain, manufactured in Great Britain by a British individual or firm, if possible from British materials. The honorary Organizing Secretary was Mr John de La Valette. The preparation took more than a year, and the amount of work done during that time was great.

The procedure of selection was as follows. First, Advisory Committees were set up, consisting of manufacturers and artists. Neither distributors nor members of the public were invited to serve, in spite of Mr de La Valette's pleadings. Their task was to interest firms and to induce them to take part in the coming exhibition, which promised to be a great social event and to carry much publicity. Moreover, since the Galleries of the Royal Academy and the wealth of the famous Institution were at the service of the exhibition, there was no rent to be paid for space. The Advisory Committees were also willing to suggest to firms which of their lines seemed artistically most commendable. A first preliminary selection took place in the autumn of 1934. Only designs were judged and not finished products. The Committees acted together with some outside experts whose names were not made public. The subsequent selection of the goods themselves was in the hands of the Executive Committee, whose members were: Ph. Connard, R.A., W. Reid Dick, R.A., Sydney Lee, R.A., Sir William Llewellyn, P.R.A., Sir E. L. Lutyens, R.A., Sir Giles Gilbert Scott, R.A., P.R.I.B.A., Sir Felix Brunner, Bt., Arthur J. Davis, A.R.A., C. Geoffrey Holme,

Col. Sir H. McMahon, John A. Milne, P.R.S.A., Tom Purvis. The Executive Committee again called in whatever outside experts they thought suitable. A last selection took place when the display of the articles was arranged. But no definite "Hanging Committee" was appointed.

The exhibition consisted of the following departments: dress materials, furnishing fabrics, carpets, ceramics, glassware, gold, silver and jewellery, leatherware, book-production, posters and commercial printing, domestic equipment, furniture, plastic mouldings. A special room was devoted to an exhibition of shop-window display, which was changed fortnightly. The decoration of the galleries was entrusted to the following architects: Maxwell Fry, H. S. Goodhart Rendel, John Grey, Sir Ambrose Heal, Oliver Hill, Chas. Holden, L. Irving, E. Maufe, O. P. Milne, The Rowley Galleries, P. A. Steynes, G. Grey Wornum. To almost all the objects shown the name of the designer was attached and—as often as possible—the retail price. This, however, could not, owing to the marked reluctance of many producers, be made a hard-and-fast rule.

So much for the aims, the organization and the arrangement of the exhibition. As to criticism provoked, by far the most serious reproach was that too much in the exhibition was expensive and too little within reach of the majority of visitors (to say nothing of the majority of the population), that too much was modish and eccentric and too little of sound and plain workaday character, too much exclusive craftsmanship and too little mass production. That is what Herbert Read meant by contrasting "individual objects" with "prototypes",[1] and Frank Pick by contrasting "beauty and grace" with "function".[2] Mr de La Valette argued, when I discussed this point with him, that a Rolls Royce is as much a product of British industry as an Austin. True, but is it not rather a matter of proportion? There was cheap pottery, there were some cheap dress materials and articles made from synthetic resins, and there were other inexpensive articles as well. But the proportion between them and costly objects was wrong. Since the exhibition was called British Art in Industry, and since its President was the then Prince of Wales, who more than once had pleaded for "fitness for purpose", "simplicity of line, form and decoration" and "working

1 *Architectural Review*, vol. lxxvii, pp. 45ff., 1935.

2 Letter to *The Times*, 9 February 1935.

for the majority instead of for the needs of a minority"[1]—the great quantity of expensive goods was a slap in the face not only to those seriously concerned with the matters in question, but also to those ignorant and well-intentioned visitors who came to find out whether the Modern Movement could after all mean something to them.

This is certainly a fair and a weighty argument against the Burlington House Exhibition. In comparison, it is less important that some of the best English designers complained of not having had their share in the building up of the exhibition,[2] and that manufacturers complained of not having known exactly who was in the end responsible for the selection. But there is one more objection which is so obviously dependent on the exclusive character of much of the exhibition that it cannot be omitted. Several times during my interviews with producers I was given to understand that articles in the exhibition had been made for publicity's sake only, and not with a view to future commercial production. I do not know how often such cases have arisen. Rarely, I expect, in pottery, dress and furnishing materials. But the very existence of such cases, partly accounted for by the basic condition of the exhibition not to show anything which had been on the market before, is a serious matter, because it is apt to invalidate some of the most important results of any such exhibition. If the visitor cannot be sure of finding the objects which he especially liked, in the ordinary, or at any rate in available, shops, and if the expert cannot be sure whether the show represents individual or industrial production, the practical utility of the enterprise is gravely affected.

However, in spite of all, one ought to say that, although the Burlington House Exhibition did not reach the standards of the Dorland Hall Exhibition in 1933, nevertheless it may well have fulfilled its prime purpose, namely to call the attention of a much wider public to a problem which that public would still not have acknowledged, had it not been put forward under the auspices of the Prince of Wales and the Royal Academy.

[1] Address delivered at the opening of the Burlington House Exhibition (*The Conquest of Ugliness*, p. 7), and Message to the Design and Industries Association, 4 November 1931 (*D.I.A. Quarterly Journal*, no. 17, January 1932).

[2] *Vide* e.g. Paul Nash, letter to *The Times*, 18 February 1935.

Yet, what such an exhibition might have been! It might have set up a milestone in the evolution of the Modern Movement in England. That opportunity was missed for a much more deep-seated reason than any so far mentioned. Mr de La Valette is proud of the fact that he and the Committees were as catholic as they could be, and that they did not favour any particular taste or style. There are three fatal errors in this which, in spite of the goodwill of the Royal Academy and the Royal Society, and in spite of the amount of organization and preparatory work done by Mr de La Valette, doomed the exhibition from the very beginning. One is that the Organizing Secretary and the Executive Committee did not want to set up standards, the second is that in fact they did set up standards, of which—and this is the third—it is impossible to approve.

The reason why it seems wrong to organize such an exhibition without avowed standards will be discussed in Part II. As to the fact that there were standards, though nobody seemed quite clear about them, how could it be otherwise? The exhibition was meant to impress upon the public the beauty of English products and upon the manufacturer the fact that English artists can create original, attractive and technically suitable designs for industrial products. But what is beauty, what is original, what attractive? No doubt what the Committees regarded as such, according to their standards. One thing about these standards was indicated clearly in many talks and articles, namely, that modern design should not enjoy any special preference. The President of the Royal Academy said at the inaugural meeting of the Advisory Committee on 8 November 1933 that "design might be traditional or modern, if it were original". The same principle was put into a more distinct form by Professor Goodhart Rendel in the book on Industrial Art, edited by Mr de La Valette and published in connection with the exhibition: "New or old in style? It will all soon be old, and neither better nor worse in itself for that." In such a remark the argument of immediate live connection with the problems, the needs and desires of our own time, is completely disregarded. An easy and good-natured conservatism seems to have set the standard, as well befitted an institution which six weeks before the opening of the exhibition gave to a Royal prince an eighteenth-century bracket-clock as a wedding present,[1] and as equally well befitted an executive committee of

[1] Cp. *Manchester Guardian*, 7 December 1934.

whose twelve members six were over sixty years old, four between fifty and sixty, and only two under fifty.

As to Mr de La Valette's own taste, some quotations from his speeches and articles may give an idea of what appeals to him and of what he is up against. He describes the first influx of post-impressionist art into England before the war as "noxious gases . . . largely the decomposition products" of a French reaction against Art Nouveau.[1] And the Central European Modern Movement which began to influence England after the war, he calls a "German outburst of mechanized art, entirely alien to the spirit of either the English or the French".[2]

But perhaps it may be thought that he alludes here to the horrors of the modernistic vogue of which Central Europe is certainly one of the chief culprits? He does not, however, for in another place he expresses with more precision what he dislikes. He speaks of the "iconoclastic fervour", of rooms looking like a "prison cell", and of "functional (oh, term of horror)".[3] "In Decorative Art", he says, "in 1933 we touched bottom", that is, the bottom "of featurelessness." And when he wants to illustrate this by referring to carpets, he evidently does not mean the crude patterns and blatant colours in commercial use, but points to "pallid colours" like "faded protoplasma",[4] which makes depressingly obvious what he thinks of. But he has not lost hope. Before his exhibition opened, he wrote: "Madame is at last wearied of the hospital-cum-factory furniture. . . . To Madame I look for salvation. . . . She has so long been a 'woman', sturdy and striving after something or other, that I believe she is about to be content once more with merely being a 'lady', with all the charm, the distinction and the good taste which that term connotes."[5]

OTHER EXHIBITIONS

Once more refraining from any conclusions, I will here only outline the sequence of exhibitions connected with our subject which were held in England during the last decade. The first consciously modern exhibition of which I know was

[1] *The Conquest of Ugliness*, p. 6. [2] *Ibid.* p. 7.

[3] Introduction to "Decorative Art", *Studio Year-Book*, p. 11, 1934.

[4] *Conquest of Ugliness*, p. 160. [5] *Decorative Art*, 1934.

the one held at Waring and Gillow's in 1928. Part of it was designed and selected by Serge Chermayeff, who had been given a completely free hand and had travelled much on the Continent in order to obtain exemplary modern articles such as pieces of furniture, textiles, fittings, glass, china, metalwork, etc. The exhibition was arranged as a series of rooms of different types, and has played a considerable part in consolidating the Modern Movement in England.

In 1929 the British Institute of Industrial Art showed at the Victoria and Albert Museum that Exhibition of Industrial Art for the Slender Purse to which I have referred earlier in this book. After that nothing decisive seems to have happened until, as a consequence of the Gorell Report, the Dorland Hall Exhibition took place in 1933. It has been said before that the chief initiative was due to the Design and Industries Association, and that this exhibition remains the best survey of modern and well-designed objects which has so far been held in England.

In 1934 a series of exhibitions followed. London saw the excellent display of photographs and models of International Modern Architecture at the new headquarters of the Royal Institute of British Architects; another and less successful, because more exclusive and more orthodoxly modern, Dorland Hall Exhibition (slightly too Corbusierish, I should say, to be right for this country), and a very instructive and attractive Exhibition of Modern Living at Whiteley's. In 1935 the only notable exhibition in London, besides that at Burlington House, was the show of articles to be obtained at Bowman's, which were selected and arranged by the Design and Industries Association. At the beginning of 1936 the Royal Institute of British Architects held an Exhibition of Everyday Things, very good and interesting so far as the building trades, domestic equipment, etc. went, but of an uneven quality in the other sections. Later in the year Heal's showed an exceedingly stimulating exhibition of modern rooms, designed by Maxwell Fry, Jack Howe, Raymond McGrath, Marcel Breuer, Christopher Nicholson, Hugh Casson, Christopher Heal and Brian O'Rorke. Early this year the Royal Academy gave a survey of British architecture in the twentieth century.

Apart from such special exhibitions, the British Industries Fair, the *Daily Mail* Ideal Homes Exhibition and the Building Exhibition have to be mentioned as recurrent features of London's exhibition life, and those who are able and willing to pick out creditable design from a multitude of horrors may go to these shows and collect information. For a public more specialized in its interests, some of the Trade Exhibitions are of great importance, above all the Radio and the Motor Exhibitions, both at Olympia, and the excellent permanent exhibition of building materials and building equipment at the Building Centre in Bond Street.

Nobody can complain of insufficient opportunities for seeing new articles. The trouble is not one of quantity but of quality. And this leads to precisely the same results as our consideration of the Royal Academy Exhibition. Unless a rigid standard of quality and of style—which is, as I shall be showing later, almost the same—is set up and ruthlessly maintained, no success of real consequence will be possible.

What is needed from our point of view is not a multiplication of fairs and general trade shows, but conscientiously and uncompromisingly selected exhibitions of the best articles at cheap or at any rate reasonable prices—articles which are on the market and can be obtained by everybody. More about this will be said in Part II.

PERIODICALS AND BOOKS

In periodicals dealing with industrial art a division must be drawn between trade and art magazines. In circulation, size, quantity, influence, trade magazines are infinitely superior. In quality of products illustrated they are incomparably inferior. I know of only one trade magazine in this country which maintains a rigid artistic standard: *Shelf Appeal* (started in 1933). It is hardly an exaggeration to estimate that 90 per cent. of the goods illustrated in trade papers are devoid of artistic value. That corresponds exactly with what our survey of trades has yielded; and how could a trade magazine adhere to a higher standard, dependent as it is on its trade association and on advertisers? It would be unjust to blame the trade papers; they have nothing to do with art.*

As to periodicals of artistic standing, the most widely diffused is certainly *The Studio*, directed by Mr Geoffrey Holme. It appears monthly, but besides the magazine, the same publishers bring out a periodical on commercial art, a *Year-Book of Decorative Art*, a *Year-Book of Modern Photography*, special numbers on various topical subjects, and other art books. The tendency in all these publications is not to be too dogmatic. Things are discussed and illustrated which may not always be up to the most fastidious standards. But their large circulation makes *The Studio* and its additional publications one of the most important instruments of propaganda in the field of modern art and design.

Its irreproachable strictness of standard, its outspokenness and its admirable thoroughness have made *The Architectural Review* the most valuable periodical on industrial art in this country. Under the editorship of Mr R. de Cronin Hastings and Mr J. M. Richards, all questions are discussed and amply illustrated which concern architecture and the architectural outlook on industrial art. The special Christmas number of 1935 was a compendium of good design in articles of everyday use with names of makers and retail prices, more valuable than a score of the usual exhibitions. The special Timber number, the special Brick number and the Christmas number of 1936 were also mines of information.

Amongst the special papers on design, *Design for To-Day* has recently been incorporated with another periodical and has lost its once leading position. An excellent new magazine of 1936, *Trend*, unfortunately lasted only for a few months. Other periodicals to be mentioned are *Decoration* and *Industrial Arts* of which the latter also began to appear in 1936. The rise of so many periodicals on applied art and industrial design is a further proof of how topical the problems discussed in this book have become during the last few years.

Alongside of these magazines six books on industrial art have come out since 1934.* Mr Geoffrey Holme was the first with his book *Industrial Design and the Future.*[1] It consists of a manifesto of some ten pages, pointing to existing troubles and desirable improvements, and a questionnaire which Mr Holme had sent to manufacturers, business men, commercial artists and architects in England and America. The answers received are published *in extenso*, and constitute the most important feature of the book. Although, naturally, they vary in accuracy, they nevertheless give a remarkable picture of the prevailing situation. All kinds of deficiencies are mentioned and remedies suggested, but no consistent solution of the problem results either from any individual answer,

[1] *The Studio*, 1934.

or from the whole of the answers taken together. Some of them I shall have to quote in Part II.

The most outstanding book on the subject appears to me to be Mr Herbert Read's *Art in Industry*.[1] As in most of his writings on art, his approach is arresting, his outlook European, and his methods carefully chosen and worked out. His aim here is not so much a direct improvement of existing conditions as the shaping of the frame of mind necessary for future improvements. He does not believe in a practicable reform, unless a new philosophical (and that means here aesthetic) creed is accepted. Hence he deals in the first part of the book with basic aesthetic laws of machine art. He shows how modern industrial art may recover for us such fundamental qualities of beauty as pure proportion, and what an important part has been played and is being played in this process by that abstract art which seems so remote from life. The second part consists of short and concise paragraphs on the conditioning factors of material and manufacturing processes in pottery, glass, metal, wood and textiles. An exceedingly interesting chapter shows that in modern mass production the real task of the creative designer lies less in the inventing of details than in the assembling of standardized parts to create new shapes of a functional as well as intuitive order, a task not much different from the architect's in building with standardized bricks, windows, etc. The third part of the book discusses colour and ornament, the fourth art education.

Mr Noel Carrington's *Design*[2] and Mr John Gloag's *Industrial Art Explained*[3] are lighter reading. Both are smaller and less abundantly illustrated than the books by Holme and Read. In Mr Carrington's book the lightness has something of champagne, Mr Gloag offers clear and refreshing water. Mr Carrington believes in the dependence of art and design on the social conditions of a period. Thus one-third of his book gives short descriptions of what design was, and of how it was embedded in society, during four periods selected from English history: The Middle Ages, the Tudor period, the Golden Age of the eighteenth century, and the Victorian Age. On the same lines and in the same brilliant way is given a sketch of the social background of art to-day. The next section deals with

[1] Faber and Faber, 1934. [2] The Twentieth Century Library, John Lane, 1934.
[3] Allen & Unwin, 1934.

Morris, the Arts and Crafts and the development following them in Germany. Then the author passes to the question of design in the Machine Age, and the position of designer, manufacturer and distributor, as it is and as it ought to be. The spheres of State activity and of individual activity are defined, and in conclusion the necessity of the artist-craftsman in the future is asserted. A temperamental and very accomplished little book, well suited for an introduction to the problem.

Mr John Gloag in his *Industrial Art Explained* also gives a short historical introduction, but it deals with the immediate antecedents of English industrial art since the middle of the nineteenth century. The stress is laid on present conditions, less from the manufacturer's than from the artist's and art lover's point of view. Mr Gloag describes how modern products such as the clock, the gramophone, the wireless, the lighting fitting, the engine, the motor-bus gradually developed from an imitation of some existing product to a form of their own. He then discusses industrial architecture and commercial art in various aspects, and questions of modern lettering and display. The small and handy volume is clearly and precisely written, a primer of industrial art. The conclusion to which Mr Gloag is led, identical with one of the main points that I shall have to make at the end of the present work, is the paramount importance of the architect as "the only designer whose training is such that the manufacturer can be certain of getting the right sort of expert advice on design".

Mr H. G. Dowling's book, called *A Survey of British Industrial Arts*,[1] was written as a "retrospective survey and acknowledgement to the many agencies that have brought about the present attitude". It deals with various trades, and enumerates and discusses the most important firms. Its optimism as to the situation to-day is remarkably opposed to the pessimism on which I felt bound to base my considerations. In his chapter on carpets, for example, to give just one instance, the author adds to the names of some six or more firms the remark that "all of them show an alert attitude for the value of good design". The volume is illustrated by a hundred plates. But since they are preceded by a note saying that "it will be obvious to the reader that many of the examples shown . . . do not conform to the principles which have been reiterated throughout the text", since, furthermore, the author leaves the reader without any help as to which

[1] F. Lewis, 1935.

of the objects illustrated he considers exemplary, and which not, and since the majority of them are in fact of the current commercial type, the value of the book as a whole is almost completely nullified.

At the time of the Burlington House Exhibition, a volume on Art in Industry was published, edited by Mr de La Valette and called *The Conquest of Ugliness*.[1] It is a collection of short introductory articles on the various trades represented in the Exhibition, surrounded by some general remarks from the editor and preceded by a contribution from the then Prince of Wales. Of Mr de La Valette's attitude, something has already been said; the Prince of Wales strongly emphasized the necessity of "simplicity" and of "substantial improvements in the design of machine-made goods, especially of mass-produced and cheap wares"; and as to the other articles, they are naturally different in character and value. I want to single out Mrs Allison Settle's interesting paper on Fashion and Fabrics, Mr H. Curwen's chapter on Perfection in Print, and Mr J. Hogan's on the Miracle of Glass. But they are by no means the only interesting parts of the book.

In conclusion I have to refer to some other methods of propaganda by means of which good work has been done for the promotion of better design. As in so many other respects, the B.B.C. have been among the first to show themselves sensitive to an imminent change for the better. From April to June 1933, that is before the opening of the Dorland Hall Exhibition, a series of talks on "Design in Daily Life" was given, the speakers being men such as Wells Coates, Maxwell Fry, John Gloag, Frank Pick, A. B. Read. The lectures were published, with a number of excellent illustrations and an introduction by Noel Carrington. In view of its date this piece of initiative is greatly to the credit of the B.B.C.

Since then many lectures and talks have been given. The Design and Industries Association, the Society of Industrial Artists and other bodies have borne their share, and it is not the least merit of the Royal Academy Exhibition that it has given so much publicity to the problem. Though the lectures which are now being delivered all over the country are naturally of varying quality, they all help to create and intensify that atmosphere of consciousness which seems nowadays indispensable for improvements.

[1] Methuen, 1935.

PART TWO

CONCLUSIONS

"Your business as manufacturer is to form the market as much as to supply it."

John Ruskin, 1859

RATIO BETWEEN GOOD & BAD IN ENGLAND & ABROAD

The previous sections of this book were bound to contain a confusing quantity of data and much criticism. It could not be otherwise. One needs a large amount of facts, in order to give a solid foundation to any suggestions for improvements, and one needs outspoken disapproval, partly for the same reason—we must know exactly how bad things are before we can take steps to alter them—and partly because any discussion dealing exclusively with the 5 or 10 per cent. of products which are well designed and well made is taking place in far too remote a sphere, a sphere which is of importance only to a small and ever decreasing class of people. To leave out the 90 per cent. of commercial goods, bought by the overwhelming majority of the nation, means to deprive the action intended of most of its vital value. But whereas fighting for improvements that concern only the "upper ten" would hardly be worth much energy, any struggle for reducing badly designed objects in general from 90 to 80 or 75 per cent. of the total production is no doubt a matter of utmost seriousness.

I said in the Introduction to this book that 90 per cent. of English industrial products are artistically objectionable. That may have sounded an exaggeration at the time. My survey of trades, I am afraid, has shown that it was a mild statement. I said also that I do not know of any country in which the majority of goods are aesthetically acceptable, but I think it is a fact that in some Continental, especially Central European, countries the ratio is more satisfactory than in England. To support this assertion I wish to quote some distinguished witnesses. Mr Frank Pick, in a speech at the Royal Society of Arts,[1] mentioned those "Continental neighbours who have given much closer attention for a long period of time to this problem of importing into industry the elements of art and good design". Herbert Read writes at the beginning of his book on *Art in Industry*: "The artistic quality of manufactured goods, especially in those countries influenced by the Bauhaus ideas of Professor Gropius, is undoubtedly higher than in Great

[1] *The Journal of the Royal Society of Arts*, p. 449, 20 March 1935.

Britain." John Gloag in *Industrial Art Explained* regards "most European countries" as "far ahead of England in those sections of industry that depend . . . upon a working partnership with design".[1] And Noel Carrington compares the wealth of England with the economic collapse of Germany after the war, and states that the German crisis "made Design and Industry partners whether they liked [it] or not", while English stability "proved an obstacle to its (the new industrial art's) quick reception".[2]

Whoever wants to see with his own eyes the truth of such depressing statements is invited to spend a morning at the British Industries Fair, an afternoon at a department store, and an evening in examining a number of trade magazines at a Public Library.

My primary aim in this second part of the present work is to summarize and tabulate the results of my research in conformity with various principles. It will be clear now to any attentive reader of the preceding sections that such results are bound to be far more complex and apparently contradictory than they might seem from most of the usual discussions on the subject. In my interviews with manufacturers, distributors and artists I could not fail to see what an alarming quantity of both sound and futile arguments, of arguments often overlapping and sometimes even excluding each other, must be considered, before any suggestions for improvements can be put forward.

SEQUENCE OF TRADES ACCORDING TO ARTISTIC QUALITY OF PRODUCTS

Looking back upon the trades discussed more or less fully, it is evident that their products differ greatly in artistic value. If I try to arrange them in an approximate order of merit of design, the first rank would be held by objects such as sports articles, travelling goods, motor-cars, sanitary appliances, watches, metal windows. Lettering, on paper as well as on metal, is hardly inferior. In furnishing fabrics, above all tweeds, linens, crashes and printed materials, some English goods are also amongst the best produced anywhere. The same is true of a few wireless cabinets, electric cooking ovens, synthetic resin products

[1] p. 25. [2] p. 72.

and some pottery. The satisfactory or even the creditable pieces would here not be in a majority, as in the goods mentioned before, but one would be able, by making a careful choice, to buy pleasant or acceptable things in most average shops—a criterion already used. This one could also do with fancy-leather articles, and, to a lesser degree, with silver and glass. The situation is worse in the wallpaper and furniture trades, where objects of good modern design are almost completely confined to a few exclusive shops. And then, in passing on to jewellery and carpets, we are at the bottom.

REASONS FOR DIFFERENCES OF ARTISTIC QUALITY

This strange sequence, which at first appears to be completely illogical, must be the outcome of explicable conditions, and, with the aid of the evidence collected, we ought now to be able to explain it. Is the higher artistic standard in some industries and the appallingly low standard in others due to the producing, the distributing, or the consuming forces in the trade process? And as far as it depends on the producer, does that mean on himself personally, on the organization of his firm, on certain methods of manufacturing, or on something else? We shall have to deal with each of these problems separately and to find out which of them may yield serviceable answers. A first group of questions concerns the manufacturer and the factory, i.e. the processes of production, the use for which goods are manufactured and their price; a second group the person responsible for the designing.

SIZE OF MANUFACTURING UNIT

There does not seem to be any tangible relation between the size of a firm and the artistic quality of its production. As to small units, they are, in my experience, usually either very good or very bad. A firm may have remained small, because the owner wanted to limit his products to a particularly good quality, or because he is just lacking in enterprise. He may be an artist-craftsman or a man of an exacting personal taste who is not willing to embark upon volume production, or simply a narrow-minded business man who has inherited the firm and wants to live comfortably on its profits without investing much money in improvements of plant or design. So he will muddle on in the same groove as his father and grandfather, as long as no serious crisis forces him into something new. The first type is to be found among the highest-class furniture-makers and studio potters, and also in a few of the best firms in other trades such as glass, textiles and electric fittings; the second is frequent among the Birmingham jewellery and light metal manufacturers, the Walsall leather firms, the High Wycombe furniture factories, the Stoke potteries.

The big mass-production firm is a priori neither better nor worse. By mass production I intend here and everywhere in this book the production of large quantities of exactly identical articles with the aid of machinery. I have seen a good many works to which this description applies, but have found it impossible to generalize as to the artistic merits of their output. It is advantageous to improvements of design that in many of these firms a progressive spirit reigns and sufficient money is available for experiments. This applies especially to the motor and wireless industries which are almost exclusively concentrated in large units. On the other hand, a big combine or mass-production firm may easily lose that personal and human element which can be so favourable to the artistic quality of the objects turned out, and be restricted to providing "what the public likes", i.e. insignificant stuff on lines which have been successful for years past and are recommended by buyers and travellers. Moreover, production on a large scale implies a reduction of the quantity of patterns or models run at the same time—the fewer the models the fewer the different manufacturing operations—and that results in a reluctance to risk experiments.

Yet if we consider the quantities turned out by mass-production firms and think of all the houses into which their goods go, all the human beings who will be surrounded by them—how depressing it is that more advantage is not taken of the great possibilities which the mass producer has for increasing harmony and order in our lives. Here is the furniture works selling 50,000 suites of one successful design in a year, here the cheap jeweller selling 200,000 cuff-links of one pattern, here the manufacturer of cheap motor-cars selling 100,000 cars a year, the firm manufacturing plastic mouldings selling 300,000 of the same tumbler each year, the aluminium works selling 100,000 kettles of the same type, the department store selling over 30,000 yards of one cheap satin in one colour during one year. What an overwhelming responsibility! Firms with such a volume of trade ought to have sufficient money and understanding to counterbalance this concentration on a few models by an equal concentration on the best available suppliers of designs, as, in fact, is the case in certain motor-car, wireless and cooking-stove factories. For a reduction in the quantity of designs by no means necessitates a reduction in their quality.

So the only result we find as to the relation between size and quality is that the large unit can be helpful by using its abundant means and progressive spirit in favour of design, and the small unit can be helpful by using the personal taste and energy of the one man controlling the production for experiments on a small scale. But in practice, the ideal representatives of both these types are rare. Most mass-production factories rely on routine work for the sake of easier money-making, and most small manufacturers are too indolent or too much concerned with securing orders and keeping afloat to recognize their social duties. So it seems, broadly speaking, that it is not the size of a firm which decides the matter, but the personal attitude of the manufacturer.

ORGANIZATION OF MANUFACTURING UNIT

The problem of the relation between size of unit and design is closely related to that of the relation between organization of unit and design. As a rule, the small unit is run by one person or a small number of persons, the big firm has a more complex and less personal form of organization.

However, there is one special question connected with this which requires a few words to itself. I was particularly anxious to see factories belonging to co-operative wholesale societies and to interview their directors. It was possible for me to look at six works in England and Scotland. One of them was a large mass-production factory with 2000 employees, the others each employed between 140 and 400. My idea was that there should be educational possibilities in co-operative enterprise which might have favourably influenced the appearance of products. This I found was a misconception. The design of the articles which

I saw—and there were pieces of furniture, cutlery, pots and pans, linoleum, packages, etc., amongst them—was in no way better than that of similar goods produced by private manufacturers or limited liability companies with no particular interest in good design.

I tried to find out why this was so and was presented with various arguments. The blame was by some of my informants laid on the customers. The bulk of them are, I was told, people with incomes ranging from £150 to £250 a year. Their artistic standards are low and uneducated. The blame was also laid on the retailer; and here is a more interesting point. Co-operative factories are only allowed to sell to co-operative shops, whereas these shops can buy from private manufacturers as well as from co-operative works. Thus the designer in the co-operative factory is even more dependent on the shopkeeper than the designer in other works. If in a small town the manager of the co-operative shop dislikes the look of a product made by the co-operative factory, say the pattern of a linoleum or the design of a package, he can desert the co-operative principle and go elsewhere to get what he likes.

This argument, however, is entirely based on the assumption that the retailer is necessarily partial to bad design and averse to good design. Are the managing directors of co-operative factories entitled to say that? In point of fact, they hardly seem to have tried to push goods of a clean modern taste. When I pointed this out, they replied that here again the co-operative factory is seriously handicapped. A private firm may risk a part of its annual profits in order to satisfy a desire for better design. A co-operative society feels, far more than any limited liability company, responsible to every one of its shareholders, and therefore does not consider itself authorized to take any risks of that kind.

But is not this looking at the matter from the wrong angle? How about factory buildings? I was struck with the rational lay-out and the modern hygienic equipment of almost all the co-operative works to which I had access. They were in most gratifying contrast to the majority of factories in this country. I wonder if these buildings are really absolutely necessary, or whether they were not put up in view of the societies' duties towards the working classes. If that is so, and if the societies are as keen on giving their customers the best possible value as they all told me, why do they not include good design amongst their recognized duties? They argue that, whereas consumers would all agree on what is the best milk or which is the most efficient bicycle, they would be greatly at variance if asked what they considered to be good design. This cannot be denied; and yet what appears to be needed is not an infallible solution of the problem, but much goodwill. If the Co-operative Movement would only start to concern itself with the problem, there would certainly be enough socialists amongst the supporters and promoters of the modern style to supply factories with competent advisers.

OLD AND NEW INDUSTRIES

I do not find myself able to make any conclusive statement on the relation between the age of an industry and the artistic quality of its products. On the whole, I should say that industries which were started in this century have reached a higher level of taste than most of the old-established industries. This would hold good of motor-cars, wireless cabinets and gramophones, refrigerators, objects in bakelite, beetle and similar materials, wrist-watches, electric fires, and, of course, aeroplanes. In all these cases there was first a period during which other objects of a superficially similar character were imitated. This was followed by a rapid development towards a truly contemporary, genuinely suitable and aesthetically satisfactory form, which as a rule was due more to engineers than to artists.

On the other hand, it would be extremely unjust to condemn root and branch all trades of an older tradition. Some of the very best pieces of English industrial art are produced by trades with a strong and live tradition. I have already mentioned Wedgwood's eighteenth-century shapes, the eighteenth-century shapes of English cutlery, the conventional and yet pleasing patterns of so many printed fabrics and, above all, the character of English lettering. All this is distinctly conservative, it is not exciting, but it shows a dignity and balance possible only in a country whose traditions have never been blocked by any revolution. The typical English attitude towards tradition and modernity may be illustrated by two quotations. The first is taken from Lethaby's writings: "We ought to recognize tradition and add something which shall be the tradition of the future." In accordance with this, Mr Frank Pick once said that industrial design in this country should be "modest and not grandiose in scale . . . not too logical in form . . . a reasonable compromise between beauty and utility, neither overstressing beauty till it degenerates into ornament, nor overstressing utility until it becomes bare and hard".[1]

But a recognition of this inherent quality of the English does not imply the imitation of period forms in English industrial art. The decisive question is whether the forms of the past are still alive or not, and the answer to this question varies according to which trade you are regarding. In lettering and pottery no drastic changes have occurred. A cup is still a cup, and an A an A. But, when you come to imitation eighteenth-century decoration applied to cups by the modern process of litho-transfer, and when you start adorning the entrance to the headquarters of a bank or an industrial combine with an elaborate gate the forms of which were invented for an eighteenth-century country seat—then the position is fundamentally different. New processes, and new purposes as well, require new forms.

OBJECTS FOR USE AND OBJECTS FOR ADORNMENT

The problem referred to just now is closely connected with that of the different artistic value of objects for use and of objects for adornment. Here a clear answer can be given. It is definitely the objects for use that are to-day more satisfactory in form. In many industries it was alarming to see that the more art is applied to an article the worse its appearance becomes. This result was obtained by comparisons between various electric fires, between various electric fittings, between watches and brooches, between steel windows and bronze gates, between beetle tumblers and celluloid mirrors. The Doulton acid jars, the Aga and Esse cookers, motor-cars of a good many makes, tubular and spherical lighting fittings, reflector fires, saddles and golf bags—they are all irreproachable, because their forms have been developed in order to satisfy practical requirements, and art in them is simply carefully considered proportion and well thought-out detail. Producer and consumer are satisfied with such

[1] Letter to the Society of Industrial Artists, published in *Studio*, April 1931.

objects as efficient machines which frankly express their own qualities and purpose. But alas for the object of industrial art which suddenly has to be artistic apart from its purpose! A perambulator is enriched by golden scrolls, an electric lamp by glass-carved satyrs or the like, and a plate by a cottage, trees and lupins.

That all this leads to such appalling consequences to-day, is by no means a matter of course, and is well worth some thought. Why are Renaissance ink-stands with a naked figure of Justice, why are eighteenth-century plates with Chinese temple scenes satisfactory? Not because there are any laws demanding or forbidding a certain amount or a certain type of decoration, but because every period creates its own style according to intrinsic laws of vitality. I had to emphasize more than once that most modern architects feel at sea as soon as they are asked to create for decoration, for adornment only (artistic metalwork, elaborate electric chandeliers, jewellery). Now we can give a reason for that, though one which I am afraid will not satisfy everybody. It is the "Zeitgeist" of our age, the age of steel, of speed, of work, or whatever one may call it, that does not allow for much play and much carefree enjoyment. One may regret living in such an age, but one cannot alter it, certainly not by imitating the outward forms of other periods.

With all this, I do not mean to assert that there are no products of contemporary industrial art which are decorated and yet up-to-date. But they are rare. And how severe is even Gordon Russell furniture, or Keith Murray pottery, or Powell glass, as compared with the festiveness which Art created in times when the lives of the ruling class were untroubled leisure and delight.

HAND-MADE AND MACHINE-MADE ARTICLES

First of all, there are no intrinsic distinctions of artistic rank between handicraft and machine art, though many distinctions of kind. Herbert Read has sufficiently enlarged on this in his book. I have seen products representative of all grades between hand-weaving, hand-thrown pottery, hand-made furniture and the manufacture of mouldings in synthetic resins, of steel windows, steel furniture, pressed steel car-bodies. Only in a few exceptional cases, however, had I to deal with craftsmen who acted as their own designers. This method, universal throughout the Middle Ages and still usual in the reigns of the four Georges, is now confined to a few jewellers and silversmiths, some cabinet-makers, studio potters, hand-weaving artists, etc. Commercially speaking, their workshops can hardly be taken as business enterprises at all. Quite often they are not paying propositions, and the artists live on "something of their own". Nevertheless the great experimental importance of such workshops is obvious, and they ought not to disappear, or at least they must have a refuge in art schools. For they alone keep up a direct unity between design and execution, and thus convey to the designer a live understanding of materials.

In all the trades which I examined, even those that depend on highly skilled labour, the designer has nothing to do with the manufacturing. The man who blows glass in Venice or cuts glass in England, the man who works in the smithy at an elaborate period gate in wrought iron or in the jewellery shop at a precious diamond brooch, may possess an admirable dexterity, but he is not an artist. The creation of shape, proportion, colour, pattern is not his, but the designer's. The work of the designer, on the other hand, is essentially the same, whether he prepares a drawing for a skilled worker or for a "hand" in a completely mechanized factory. Great changes have taken place, and sad changes, no doubt, between the days when manufacturers still "manu faciebant", and our century, when the "hand" has become the slighting expression for an entirely unskilled operative. However, we cannot put the clock back, although some countries try to. Had these changes not occurred everywhere, the problem of the designer, the problem of industrial art as opposed to "les arts décoratifs", would not exist.

As to the aesthetic value of articles produced under present-day conditions, this is likewise bound to be different from that of past centuries. The charm of a piece of medieval handicraft lies to a large extent in its being the live expression of something human. No contrast of attitude existed between the goldsmith chasing a chalice and the painter painting an altar. The picture was not pure art, and the chalice not pure handicraft. Either represented an indivisible ("individual") unity of soul, mind and hand. The situation changed as soon as the artist began to think of repetition, no matter whether in pottery or the graphic arts. Even a fifteenth-century woodcut depended as a rule on two separate personalities, the artist who invented the composition and drew it out, and the cutter who transferred it to the wood-block. Only one of them needed to have artistic inspiration, the other's natural gift was to be skill, appreciation, sensibility, but not the genius of art. The final effect certainly always depends on both, in that deficiencies on either side can spoil the whole. But while no limit to artistic genius exists, the craftsman's limit is reached when he has rendered the designer's intention completely. Now in many trades the machine is infallible in doing that, if the designer knows how to design for the machine. Hence it is just pretending if one insists that a hand-cut leather bag must be aesthetically better than a machine-cut one, or hand-thrown vases better than machine-thrown. This was true only so long as designer and craftsman were one. Incidentally, it is one of the reasons for the low artistic standard of some trades which use little machinery, such as the jewellery trade, that in relying entirely upon skilled craftsmanship and doing without proper designers, they mistake technical dexterity for creative imagination.

All these considerations show that neither machinery nor repetition-work in itself is detrimental to the aesthetic quality of a product. On the contrary, there is one great possibility in mass-production methods of which advantage is seldom taken. I pointed out earlier that a firm turning out very large quantities of some few objects will necessarily tend to reduce the number of different models, in order to simplify production processes. By concentrating upon a few designs more money can be spent on each of them, higher fees for artists should be available and better men can be obtained. This has in fact been the policy of some wireless factories, some motor-car works, and a few other firms. I say the policy, for it is a matter of deliberately chosen policy and not the necessary outcome of a certain method of production. It depends once more on the personal attitude of the manufacturer. One instance in proof of this was the American carpet manufacturer mentioned on page 62 who, although running only about twenty stock patterns at a time, seemed nevertheless not to have improved on their standards of design at all.

If I say that standardization in itself is not of disadvantage to art, I do not deny that mechanical methods of production sometimes may restrict or even exclude certain improvements of design. An artist who creates new models for a large furniture firm must, in the proportions which he chooses, be controlled by the standard sizes of mechanically produced units, such as legs, drawers and backs. A designer for cheap jewellery must, in designing a new clip or bracelet, make use of existing stampings. A designer for roller-printed fabrics must not exceed a certain size of repeat and a certain number of colours. A designer for cheap car-bodies has to consider available accessories. A designer for bakelite must take into account what shapes can be easily moulded. These are only a few instances of what one comes across in almost every trade. But whereas the artist, in all such cases, is undoubtedly less free than his colleague designing for hand-production, the consequence, if he is good at his job, is by no means a reduction of aesthetic values. Every architect is just as dependent on by-laws and on the standard sizes of bricks, window panes, etc. His artistic genius comes out in reconciling awkward circumstances and creating out of them, and in spite of them, a new and personal unity. It is one of the most brilliant conceptions in Herbert Read's book, which I have to quote so frequently, that he stresses the fundamental identity of the architect's and the designer's task in the Machine Age.

While the conditions discussed so far limit only the range and not the value of the design, there is another limitation imposed on design by machinery which is indeed a more serious obstacle to artistic improvements. I said in the section on carpets that the cost of a new set of Jacquard cards or that of harnessing a loom for a new Axminster pattern is astonishingly high. The same is true of the engraving of rollers for textile printing, of the making of tools for articles in synthetic resins, aluminium, enamel, etc., of the printing of a complete edition of lithographs for a cheap pottery pattern. In all such cases, a considerable sum has to be invested before production can begin. Naturally the consequence of this is that the manufacturer shrinks from experiments. As long as he "gets away" with his goods, any change seems a foolish risk. He will not take such a risk, unless he is prepared to take it for the benefit of the community. It is easier to embark upon a new style in hand-tufted rugs, hand-printed textiles, hand-mounted jewellery, hand-thrown pottery, blown glass, hand-beaten silver, where a failure means no loss of capital, but just the loss of time and material spent on the one article or on a small range. In view of the figures which I gave before—an Axminster carpet pattern being a failure unless at least 1000 square yards are sold, a pattern for lithographed pottery unless at least 20,000 pieces are sold, a wireless cabinet in bakelite unless at least 10,000 are sold—one cannot condemn an industrialist too severely because he does not jump at every suggested artistic improvement. It would be absurd to suggest to the producer that he ought to ruin himself for the community, but what one is justified in asking is that he should consider carefully and conscientiously how he can best obtain a balance between his business and his public duties.

BOOM AND SLUMP

I said just now that as long as firms are able to keep their machines steadily going, they will hardly be inclined to consider drastic changes of design. Consequently a trade boom seems detrimental, a trade slump favourable, to the artistic quality of the products of a nation. It is not for me to go into the basic premisses of this bewildering phenomenon. Probably we are here faced with one more outcome of the uncontrolled development of private enterprise in the nineteenth century. In any case a sufficient number of instances can be given in evidence. The United States was regarded as extremely reactionary in taste until the crisis of 1929–30 brought a change. The design of American cars, for example, was far behind European standards up to that time; afterwards it took the lead, when the vogue of streamlining was created in order to induce people to acquire new cars. England remained reactionary for many years after the war. We saw that most improvements which are noticeable now have come into being during the last three or four years, that is during the time when Britain was involved in the world crisis. Germany became the foremost exponent of the Modern Movement at the time when her industries were suffering most acutely from the consequences of the war and Versailles. The same, to a certain extent, holds good of other Continental countries. This was realized by the English Board of Education Inspectors who were sent over in 1932 and 1934 to collect evidence on Continental trade schools and schools of arts and crafts. Mr A. Abbott says in his report that in Holland "the pressure of circumstances has produced a widespread recognition of the vital importance" of sound trade education, and

Mr E. M. O'R. Dickey and Mr W. M. Keesey make the same point more than once in their report.

I do not want to be misunderstood. There is no natural law forbidding improvements of design in times of flourishing trade. If there is a demand for decent standards in a nation, or if the moral energy of a group of manufacturers is strong enough, good industrial art can be produced at any time. That is shown by Germany during the years before the war, by Sweden during the years immediately after the war, and by England at the present time if one looks at such booming trades as motor-cars and wireless cabinets. What I want to emphasize here is only that, taking the normal human being and especially the normal manufacturer, the stimulus needed for a serious consideration of the importance of design is not strong enough to lead anywhere as long as business is flourishing. It is obvious that exchanging new tools for old, new rollers for old, new stampings for old, means stopping production for a certain time, and slackening it for a still longer period, even if there is not also a loss of sales after the change. Against all such understandable hesitations stands the one moral argument that it is unpardonable consciously to offer for sale goods which are badly designed. There is no doubt towards which side the pendulum usually swings.

Things are different during a slump. The manufacturer sees that his business is decreasing. Machines have to be stopped, shed after shed lies idle. This is a moment for speculating on new methods and new incentives. One of them is the appearance, shape, decoration, colour of the product. At last this is seriously examined. The manufacturer says to himself: Suppose I try something new on a couple of machines. They are idle in any case. Perhaps the public will like the new line and buy large quantities. It may even be worth communicating with a first-rate designer or artist whose genius or whose name may be of some help.

And while for such reasons he tries to improve the aesthetic merits of his goods, some other manufacturers in the same distressed industry will have had similar ideas. Artistic competition arises, and will in its turn lead to further and intensified efforts.

THE CREATOR OF DESIGNS

So far most of our conclusions have had of necessity to be rather vague. We are on safer ground in considering the relative aesthetic qualities of industrial products designed by staff designers, free-lance designers, artists and architects, engineers, managers and directors.

It is a fact which has been mentioned at various times that in small firms as well as in certain mass-production firms no full-time designers are employed, because there would not be enough work for them. In factories turning out aluminium, enamel or bakelite goods, wireless cabinets, motor-cars, etc., the small quantity of new models produced each year makes a creative staff designer appear unnecessary. In most of such factories designs "just happen", based on vague suggestions from salesmen, managers, foremen, etc. In some other works it is regarded as necessary for the sake of dignity to have a well-paid designer, although his activity is but limited; in a few firms different and preferable arrangements have been made. These will be discussed later.

Another reason for not employing staff designers is the small volume of trade in some industries, such as pottery, glass, jewellery, fancy-leather. Here again designs either "happen" in the way described before, or else they are entrusted to inadequately trained, ill-paid men taken from the bench and put in charge of what is called the design department. In fact designs are as a rule worse where a so-called designer exists than where new models are produced without the interference of any pseudo-artist.

We have found a great variety in the salaries of staff designers. The small or medium-size factory producing pottery, glass, jewellery, door furniture, electric fittings, furniture, textiles, pays anything from £150 to £450 a year, say £250 to £300 on an average. In jewellery I know of no designer earning more than £450, in pottery there are a few head designers in ambitious firms getting anything up to £750. In the furniture, cotton, carpet, wallpaper and motor-car trades, an extraordinarily good man, working in one of the largest firms, may be paid anything between £500 and £1200, or even up to £1500 and more. But such cases are extremely rare, some two, three or at most five in each industry. No young designer leaving an art school can count on ever reaching a creditable salary. This is what so often deters able art students from specialising in design. In fact, it is not even artistic talent which usually determines the income of a staff designer. The size of his firm is far more important. An excellent pottery firm, requiring a man of the highest imaginative qualities, may be compelled to pay less than a big mass-production works requiring his services for the design of steel windows.

Bad payment is one of the causes of the poor quality of so much work done by staff designers. Another is to be found in their conditions of life and work. Most of those whom I saw have to spend their lives in small provincial towns such as Stoke, Kidderminster, High Wycombe, Stourbridge, Halifax, etc. They live in dreary streets, work in bleak studios, and have no chance of travelling to refresh their imagination. It is depressingly characteristic that the Board of Education Inspectors reporting on the conditions of design in the cotton industry in 1929[1] denied that it would be much good to transplant a designer from Paris to an English industrial town. The reason given was that "if he is transplanted to England and particularly to Manchester, he loses his freshness in a short time". Here we are obviously faced with a serious social question. Its bearing on our problem will be indicated in a separate section.

A third and more special reason for the lack of artistic refinement in the work of most staff designers is the continuous interference of sales departments in many factories. Nobody would object to the interference of engineers who tell the designer what is technically necessary and advisable. Nor would anybody object to sales managers communicating to designers which of the year's new patterns or new models were successful and which not. But when the sales expert tries to indicate what ought to be done in future, the result is as a rule aesthetically disastrous. There is nothing astonishing in that. A layman may well tell the painter that he does not like a certain combination of colours, but there would not be many laymen who would venture to suggest to the painter what he should do to get a better colour scheme. Accordingly the sales manager will usually be good at negative suggestions and unqualified to give anything more than general positive suggestions.

Some arguments against the exclusive use of the staff designer have been put

[1] Board of Education, Educational Pamphlets No. 75, p. 10.

forward. The next question is, how can he be replaced? In some industries, particularly carpets, printed fabrics, and wallpapers, he is to a certain extent ousted by free-lance designers (about 25–50 per cent. of the designs in carpets, about 40–60 per cent. in cotton printing are bought from outside). It is a characteristic fact that this has become usual only in trades where designing means surface decoration and not the shaping of the article itself. The invention of patterns can be done easily in a studio remote from the workshops; to create a new shape for pottery, glass, weaves, bakelite, plate, etc., the designer has to be near the place where the objects are produced. I said that many of the "Public Designers" are foreigners, that they are preferred because their imagination is regarded as fresher and less "tradey", and that payment varies between £2 and £10 a sketch. For finished drawings higher prices are often paid, which shows once more that it is not artistic quality alone that decides the amount of money spent on a design. But as a rule sketches are more popular. The average fee for one of these is about £5. In fancy leather, where the use of free-lance designers is less widespread, £3 is regarded as a very good fee. But in textile printing and weaving a well-known artist may be paid up to £35 and £40. Hence it is not bad payment that is an obstacle to the success of this system. And yet we have seen how appallingly bad is the standard of design in carpets and wallpapers. The situation in these two industries tells strongly against the exclusive use of the free-lance designer, although we have seen that this failure is not entirely his fault. The worst crime is committed in the works' studio, where the sketch bought and paid for is re-drawn and altered. I need not describe again the violations that I saw with my own eyes, pretty sketches almost killed by brutal assaults of unfeeling staff draughtsmen. But the financial position of the free-lance designer—with the exception of a very few commercial artists in London—does not allow him to object. Once having sold his child, he has forfeited any right to interfere with what is inflicted upon it in the factory.

Thus it will be seen that the two most universally accepted systems for supplying designs are by no means satisfactory. We must examine other methods to find out what can be recommended. What are the results likely to be if the manufacturer himself, the managing director, or the works' manager acts as creative designer? It seems to me that they are surprisingly good, provided the director or manufacturer is sufficiently interested and well educated and not devoid of inborn sensibility. This is the case in many car factories, and I found it also in two of the best English factories producing electric fittings, one factory producing some remarkably well-designed aluminium goods, one steel-furniture factory, etc. On the other hand, quite the contrary result is obtained where the manufacturer or manager sketches out new lines of production, because he does not consider the matter important enough to entrust it to an expert, or because he wants to save the expense of a designer. I have a right to say this after having seen so many bad articles which the manufacturer told me had been designed "between myself and the foreman", or as the foreman would say, "between myself and the boss". By emphasizing the fact that a manufacturer can do excellent work without the aid of a full-time designer on the staff, I mean to say that it rests with him to improve the value of his wares, and that it is his responsibility if he does not. But then he must realize the seriousness of the problem, must train his appreciation and deepen his consciousness of beauty in his house, in his office, in the articles of his personal wear and use.

However, artistic gifts, creative as well as appreciative, are inborn. So this method can be applied only in a small number of cases. All those firms in which the director, or manager, or owner, is not able to take charge of the creative designing himself, have to think of other methods. In some trades the chief engineer or another engineer of high standing has proved very good as a designer. I refer to the Portcullis gas fires, the Ferranti reflecting fires, the Aga cookers. The lesson from this may be that for objects dependent on technical conditions a conscientious and ambitious technician may produce cleaner and more satisfactory designs than a designer who wants to do something artistic.

In the many industries where this system would not work, I should recommend as the best method the employment of a part-time designer or artist of high standing. A small firm would pay as much as before, say £200 or £300, but pay it to a first-rate man for one or two months' work, which would be sufficient to cope with the annual demand for new designs in most firms. It would then be necessary only to have on the staff skilled draughtsmen for mechanical reproduction work. Any mass-production firm could do the same. Under this system

£100 each or more could easily be paid for the few new models or patterns required and even that would be very little if compared with the amount of money spent on design by modern-minded firms in the United States, where the most popular "stylists" get fees ranging from $500 to $25,000, and may charge $100 for one day's work.[1] The designer or artist, in his turn, could under this system live in London instead of in places such as Burslem or Kidderminster. He could work for various firms in different trades, so that in the end his salary might be very good indeed. I am not blind to the fact that a universal adoption of this method would lead to growing unemployment among designers. But we have to be cruel. Bad design cannot be abolished without abolishing bad designers. Nobody would suggest that worthless pictures should be bought in order to keep mediocre painters alive. Furthering the good always implies fighting the bad.

The men to replace staff artists as designers for industry may be artists, commercial artists or architects. As to the employment of painters, the case of Edward Hald and Simon Gate at Orrefors is in itself sufficient to prove what admirable results this can give. I also found the designs done by a distinguished London painter for one of the biggest Manchester firms very satisfactory. Among the painters who worked for Messrs Foley China several produced excellent, and at the same time commercially sound designs (Ben Nicholson, for instance). It is hardly necessary to emphasize again that an artist to be employed more than occasionally by firms must have a thorough knowledge of the material for which he designs and of the processes of production. Art schools are as a rule not well enough equipped to give him this. Hence the artist must go to the works to study material and machinery before he can start designing. But it is certainly an exaggeration to say that the necessary amount of knowledge could only be acquired by a man on the staff or by a complete trade training at the bench. In some industries this may necessitate a preparatory stay of some months at the factory, in most trades even less. This view, which I formed in going round factories, was confirmed by what Mr Voysey (who ought to know) told me, and by what Mr Chermayeff (who also ought to know) wrote in his answer to the *Studio* questionnaire.

[1] Cp. G. Holme, *Industrial Design and the Future*, p. 41.

On the other hand, the manufacturer may be right in saying that he finds it difficult to deal with painters or sculptors, or to have them about the factory, because they are unbusinesslike and touchy. This is so, I am afraid, partly because, being artists, they must be like that and partly because, having been brought up with the idea of the great message of free and independent art, they want to be like that. Since about 1800 (Schiller, Keats[1]), the artist has begun to consider himself as a superior being to whom the ordinary person ought to look up. His task is to express his own infinitely important and interesting experiences and thoughts. We still suffer badly from this morbid conception. The medieval painter was a craftsman. As such he would not have thought of objecting to the carrying out of orders. Holbein designed cups and ephemeral decorations, and he painted the sign for a schoolmaster as well as he painted his portraits. We may have gained much by that liberation of the artist which started during the Renaissance, but in the end this irresistible development has proved fatal. Without it we should not have had Rembrandt, but neither should we have had the unbearable oversupply of artists whom nobody wants and whose private feelings are totally uninteresting in a century such as this. And we should also not suffer from the equally disastrous lack of good designers. This consideration determines the attitude which the socially conscious critic ought to adopt to-day. If the painter is not so great as to express with the utmost intensity and vitality the essence of his time, let him starve, unless he is willing to find the way back to serving the community.

Take some of the best commercial artists in London. They know their job and design posters, packages and catalogues, the social value of which is higher than that of any good portrait or landscape which they might paint and exhibit, while their aesthetic value is certainly not lower. More firms ought to take advantage of these artists, not only for advertising but also for industrial design. I feel sure that they would be successful, both commercially and artistically and, in any case, manufacturers would find them more practical and more businesslike than most painters.

This is also one of the advantages in using architects as designers, and only one of many advantages; for here at last is the method which I wish to recommend

[1] "Everything is spoilt by use."

above all others. John Gloag is right in saying that the architect ought to become "*the* expert on design in the community".[1] He ought to, owing to his training, which is far more comprehensive and practical than a painter's, and owing to the artistic method which he has to adopt in his work, namely building with more or less standardized parts new, attractive and serviceable units to clients' wishes. A designer, like an architect, is not a free artist. Both must believe in the moral value of serving—serving clearly defined purposes in their individual works, and serving the community in the whole of their activity. In fact, the use of architects as designers has yielded excellent results in almost all the cases which I have encountered or of which I know. The most outstanding instance is the German Modern Movement before the war. Most of those who created the new style were practising architects, though—very characteristic of that time of transition—several of them had begun as painters and then deliberately changed over to design, before they started architectural work. This is true of Peter Behrens, Bruno Paul, Riemerschmid, van de Velde, and of many others.

The new German style in furniture, wallpapers, books and many other objects of applied art, which originated about 1902–6 and was fully developed by 1907, when the German Werkbund was founded, is the work of architects. So far as English experiments in using architects as designers are concerned, I need but refer to Voysey's fabrics, wallpapers, etc. about 1890–1900, to Keith Murray's glass and pottery, to R. Y. Goodden's wallpapers, to the Plan Furniture, to Messrs Ekco's wireless cabinets designed by Chermayeff and Wells Coates, or to the London fancy-leather firm mentioned in Part I which uses an Austrian architect for designing.

As to financial arrangements in these cases, I will sum up what has already been stated. Mr Voysey received fees of £10–£12 for a wallpaper or cretonne design, Mr Serge Chermayeff received about £50 for each of his designs for radio sets, Mr Keith Murray receives the value of one or two months' work from each of the firms for which he works, and royalties on top of that. There is no common agreement as to whether royalties are a commendable system of payment. In the answers to Mr Holme's questionnaire we find some manufacturers in

[1] *Industrial Art Explained*, p. 179.

favour and some against it.[1] But that is a secondary question. Our main conclusion is that, on the whole, the use of a permanent staff designer and the use of a "Public Designer" do not work well. Instead of that, manufacturers ought as often as possible to make arrangements with first-rate architects or suitable commercial artists, if the creative work cannot be left with a chief engineer, or the manufacturer himself is not able to act as his own designer.

THE AESTHETIC VALUE OF CHEAP AND EXPENSIVE ARTICLES

It has been emphasized all through this book that the aesthetic value of the inexpensive product is the question that really matters. It is therefore now necessary to re-examine the results in various trades with regard to the relation of price and design. Looking at our problem from this angle, it cannot be denied that the artistic standard of expensive and medium-price wares is as a rule higher than that of cheap popular products, and that the improvements which we had to note referred mostly to higher-price goods. This being so it appears necessary to repeat the few exceptions to that rule which I have come across.

In recent years several department stores have introduced a series of well-designed, simple and very inexpensive pieces of furniture, such as wardrobes, chests of drawers, etc. made of natural beech-wood and a light bastard mahogany. Their construction is of course not of the best, but their appearance is satisfactory. There is also some cheap and unadorned pottery on the market which is excellent in clay and firing and creditable in design. Further, the glass counters at Woolworth's are better in taste than the glass departments of expensive shops; some cheap jewellery, chains, bangles, etc. at Woolworth's are more pleasing and at any rate more up-to-date than most of the corresponding things in West End shops; and the "Beetle" products or the "pastelware" enamel goods at Woolworth's are pleasant both in shape and colour. On the whole it must be said that, in looking round one or other of the principal Woolworth shops, one

[1] In favour, Mr J. Wedgwood and, in a more conditional way, Mr R. D. Best and Mr F. J. Donald. Wholly in favour also, Mr J. A. Milne in *The Conquest of Ugliness*, p. 27.

is surprised by a good many well-designed simple things, whether bought in order to favour good modern design or for other reasons.

There are thus a few instances where the least costly articles in a trade are particularly commendable; rather more can be given where medium-price goods are more progressive than expensive goods. This was true of electric fires, the favourite in the high-price range being the flickering type, in the medium-price range the reflector type; of high-grade porcelain, as compared with such earthenware as Wedgwood's; and of the most expensive jewellery, bronze gates, or Wilton carpets. In most other trades, however, such as furniture, textiles, wallpapers, wireless cabinets, fancy-leather goods, the best designed articles are among the most expensive.

It must be one of the most essential tasks of any research such as the present to discover some reasons for this state of affairs. First of all: Reasons which caused a similar situation in the Baroque and Rococo periods do not apply to conditions to-day. In the seventeenth and eighteenth centuries, richness of decoration and elaborateness were among the distinctive qualities of artistic expression, because they harmonized with the ideals and the lives of those classes for which most art was created. Consequently a beautiful piece of applied art was bound to be expensive, because its precious material and the quantity of labour, i.e. of highest handicraft, embedded in it, were constituent parts of that social display which all art signified. To-day even the most wealthy art-lover is a worker; the carefree life of the amateur and dilettante of the *ancien régime* has gone. Therefore, a style of our age must be an unexclusive style, and its merits must be collective merits not distinguishing one individual or one class. It shows itself at its best in factories, office buildings, blocks of flats, estates of small houses, and buildings for public welfare such as schools, hospitals, swimming pools, etc. One of the reasons why England has been late in adopting this international style is the fact that more contrasts between classes are still in existence in this country than in those which are leading in the Modern Movement. The distance between the wealthy and the poor is larger in England than in Central Europe; the rich house looking richer and the poor house poorer, the clothes of the rich more perfect and of the poor more ragged. This is certainly one of the main obstacles to improvements in English industrial art. Unless a further levelling of social differences takes place in this country, no steady development towards the aims of the Modern Movement is possible.

As things are at present in England, a strangely paradoxical situation prevails. The modern style is characterized by qualities which are essentially mass qualities. In spite of that, most of its products are only within reach of a small wealthy class. Lethaby once said that "one cannot sufficiently hide away from the ugliness of things under £4000 a year".[1] It may no longer be as bad as all that; one may substitute £1200 or £1500 for £4000—in any case, the tragedy remains that the education which usually prepares for incomes of that range is generally so traditional or "tradition-minded" that those who can afford well-designed modern works prefer genuine antiques or period reproduction. Questions of education and social questions, however, have to be left to another section. The important point in our present context is, whether there are technical reasons for restricting modern form to expensive articles. To that no general answer can be given. There are cases where the plain surfaces of contemporary style mean an increase in manufacturing costs, and there are cases where the reverse is true. I had to report that in furniture the unadorned surface requires better material and a better finish, that the same is true of bakelite, beetle, etc., where a flat surface is more difficult to mould than a rounded one, that in aluminium straight planes have to be made of a stronger and therefore more expensive material than curved surfaces, and that in glass the simple modern cut would show all those blemishes which a richer crystal-cut conceals. On the other hand, a modern carpet with few colours and a simple pattern costs less than an elaborate period or modernistic one, and in cheap pottery, based on the litho-transfer process, the lithographs would at least cost no more were they of the most perfect modern design.

Passing on to some more fundamental economic considerations we find again that generalization is impossible. Nobody can blame a manufacturer if he says that mass production is not a field for experiment. If you want to find out whether a new line will "go", you will not do so by producing thousands of it at once, but rather by starting production on a small scale. Small-scale production is as a rule costly (allowing for exceptions such as hand-made pottery), and so

[1] *Form in Civilization*, p. 220, 1922.

it happens that something especially progressive usually appears first in expensive shops. This seems unassailable as an argument; more so than a related one which I had to mention above, in discussing the effects of trade boom and trade slump on design. I said that any new departure means a financial loss to the producer at the moment when the change takes place. The larger the scale the greater the loss. Hence such changes will be least likely in cheap mass-produced wares; for most manufacturers prefer the comfort of safety to the excitement of venture. However, big business success depends in the end on taking risks and not upon untroubled safety, so that it is doubtful whether the risk of introducing new design may not sometimes be amongst the risks necessary for big success.

More decisive than all the reasons given so far are the two which follow. In the section on furniture, I quoted the statement made to me by the managers of two shops selling high-class modern objects. Both said they did not want cheap goods of good modern design, because that "would upset their market". Their customers, they said, come to them because they want "something that is different". The fact that well-designed modern things are in a minority is what makes such customers like them. This argument was put forward only twice, but undoubtedly it is at the bottom of the business policy of many shops and consequently of many producers also. I need not enlarge on how despicable it is. While enjoying your turkey at Christmas you would be indignant if anybody attributed your pleasure to the fact that the poor cannot afford to buy turkeys. You do exactly the same if you buy beautiful furniture only because it is in a better taste than that which the majority of the people can obtain.

But, the distributor and most manufacturers will retort, the argument is wrongly put. It is not as if the majority of the people would buy the better things, even if they could afford to do so. They would not, because they would not like them; and they would not like them, because their taste is thoroughly bad. We, manufacturers and retailers, have to serve the public, and that implies not offering them things that are too good for them.

An answer to this objection requires a section of its own.

THE TASTE OF THE PUBLIC

It would be absurd to deny that there are marked differences between the taste of one group and another, one nation and another. I mentioned several times the experience of manufacturers as to contrasts between London and the Midlands, England and Scotland, Great Britain and the Continental countries, Europe and South America. I referred to the fact that Oxford is supposed to prefer Oriental to modern carpets, Scotland to prefer heavier and more practical leather handbags, a stronger and better quality of linoleum, and a more ornate and solid-looking type of pottery. Generally speaking, most Continental countries go in for brighter and gayer things than England, London goes in for smarter, lighter and more delicate-looking things than the North (in furnishing fabrics, Manchester asks for heavy weight and heavy colour, London for rayon and pastel shades), and Edinburgh for brighter and more showy things than Glasgow. While within Britain a tendency is evident by which new fashions are developed in London and then gradually move into the Provinces and up to the North, it cannot be said in the same way that new artistic styles always come from the Continent into England. For one thing, to speak of the last centuries only, the Gothic Revival and the Morris Movement originated in England; and then there are Continental modes of expression to which England seems almost entirely immune. During the seventeenth and eighteenth centuries, hardly any excessive Baroque or Rococo forms were used in England, while they were predominant in Italy, Germany and, up to a point, in France also. When Paris from 1860, and Berlin from 1875, began to revel in neo-Baroque ("Wilhelmian" style), in London Norman Shaw and his friends erected well and soberly designed houses in a modernized Queen Anne and Georgian style. This English sobriety seems unalterable, and in fact so essentially English that it would be a crime to alter it. Britain's greatness appears inseparable from Britain's conservatism. It is chiefly "cunctando" that progress is made in this country. Experiments and ensuing failures are instinctively left to other nations. However, while this basic English quality accounts for a certain hesitation in the adoption of the Modern Movement, the modern style in its entirety cannot be one of those which England will in the long run refuse. For it is a simple and rational style, and moreover a style

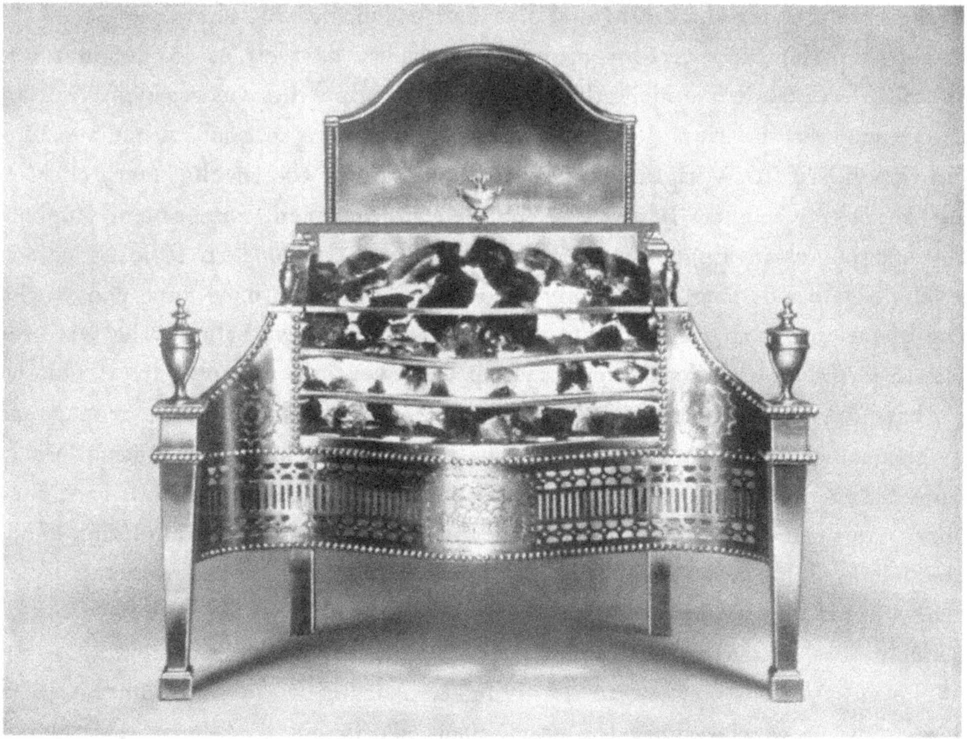

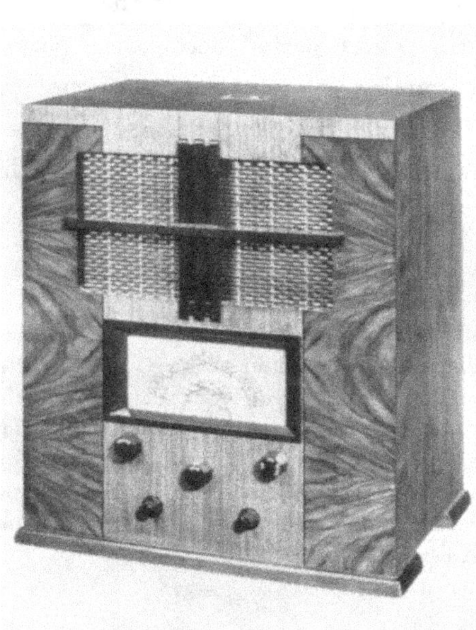

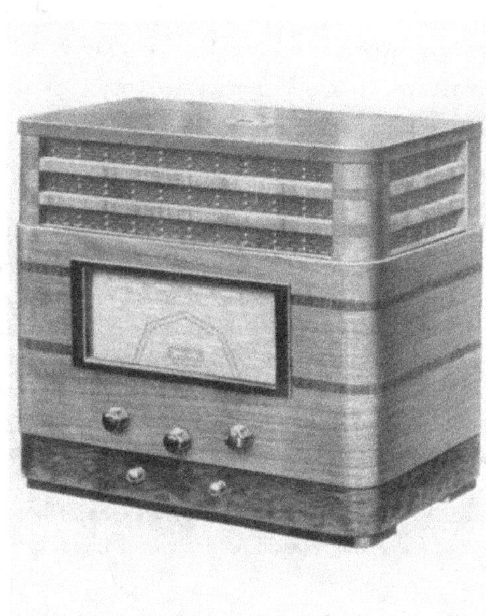

XXIII. *Best selling type of high-price electric fire (the "Magicoal" made by Berry Electric Co.).*
Two best selling wireless cabinets of 1935–6 design, made by His Master's Voice and by Marconiphone Ltd.

in the creation of which England has had a considerable share. Most of the pioneers of the Modern Movement, in Germany as well as in Belgium and America, have made it sufficiently clear that what they did was based on William Morris and his theories. It is outside my subject here to quote what men like van de Velde, F. L. Wright and Gropius have said of the revolutionary effect of the Morris Movement.[1] In any case the Modern Movement cannot be un-English, if it originated in England. In finally adopting the modern style as it was developed abroad, England is only recovering what she once gave the world. The historian has therefore carefully to collect the names of those who were the first to introduce the Continental style of the twentieth century into England.

Roger Fry's exhibition at the Grafton Galleries and his Cézanne propaganda are as memorable for English painting as is the foundation of his Omega Workshops for modern English handicraft. The activity of the workshops continued from 1912 until 1918.[2] The next memorable date is that of the foundation of the Design and Industries Association, 1915. Since then it has taken about fifteen years to create a favourable atmosphere for the modern style, and even now it is much more controversial than in some Continental countries.

The most serious obstacle to its divulgation lies in the attitude of the upper class. Owing to the general conservatism mentioned, to inborn reserve and a distrust of anything that looks strikingly new, the majority of the English upper classes, above all the aristocracy, still prefer period decoration, period furniture, period porcelain, etc.—whether genuine or reproduced—to modern industrial art.[3] In fact, I had more than once to refer to a new increase of period reproduction which has been noticed by manufacturers and retailers within the

[1] Cf. *Pioneers of the Modern Movement*, mainly chapter I.

[2] An exhibition of work done by the Omega Workshops was shown at the Bethnal Green Museum in 1934. The objects came from the Margaret Bulley Collection.

[3] This came out most alarmingly on the occasion of the wedding presents given to the Duke and Duchess of Kent in 1934, and to the Duke and Duchess of Gloucester in 1935. Cf. *Manchester Guardian* 7 December 1934: "The City of London Corporation gave Regency silverware, the London Chamber of Commerce an antique table and chairs, the Cloth Workers' Company eighteenth-century silver, Westminster City an antique dessert service, Edinburgh antique furniture, and so on." The Jubilee Presentation House which was given to King George V by the Royal Warrant Holders' Society was, according to the architect's own words, "a happy combination of Georgian, classic, and modern ideas" (*The Times*, 22 January 1935).

last year. The reason for this was revealed by a headline in a recent issue of a furnishing trade paper: "West End Trade buys Period Design. Modern schemes now too popular." This smacks badly of the "Something-that-is-different" attitude, but it may also mean that modernism in its jazz forms has spoilt the market for serious modern work.*

If that is so the consequences will be disastrous. Snobbery—the wish to imitate, if not in fact, at least in outward appearance, an envied class of which you read in the Court and Society column of your morning paper—snobbery could be a great help to the growing Modern Movement in England, if only more members of the upper class would give up Chippendale for modern furniture of equally high craftsmanship and perfect design, and old Chelsea for Keith Murray. At present it appears to be the professional class mainly, and a small minority of wealthy merchants and industrialists who uphold the modern style. Still, their taste is also bound to filter down by degrees into the semi-detached houses of the poorer middle classes, and the question now is whether these are prepared to accept the new simplicity. Most manufacturers and distributors say that their predilection for the bad, the meretricious and the showy is ineradicable. Time and again I asked them how they could know. It is only very rarely that they can. Manufacturers—to repeat what I have pointed out with regard to several individual trades in Part I—depend largely on suggestions from their own sales departments. These are based on information received from the firms' travellers who come home and state what has been a "flop" or a "washout", and what is selling like "hot cakes" or like "a house on fire" or like "ripe cherries", according to the traveller's favourite metaphor. But is this really an indication of the popular taste? The traveller in his turn relies on statements which he hears from the buyers of department stores or shops, and from shopkeepers. And as to the shopkeeper, he unconsciously pushes what he likes and conceals in dark corners what seems to him unattractive. He may also kill articles by bad and overcrowded display, let alone the fact that he may have his special and objectionable reasons against good modern design in cheap merchandise. As to buyers and travellers, they are usually not in the best position for energetic action in favour of anything new. A traveller paid commission on sales transacted, a buyer in a store entirely dependent on turnover, these men

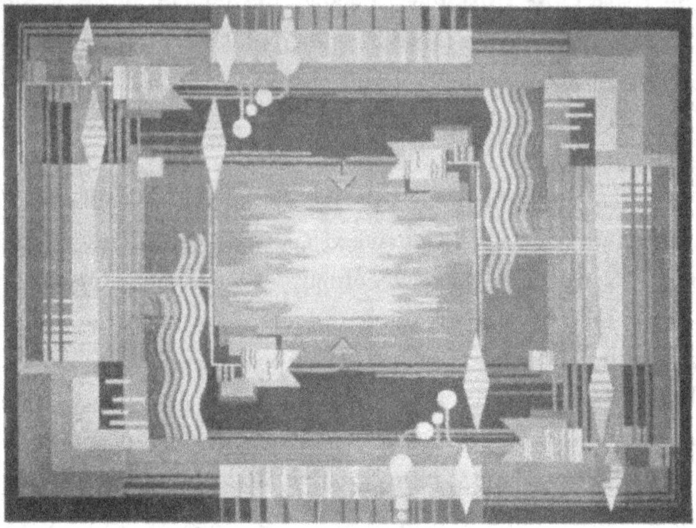

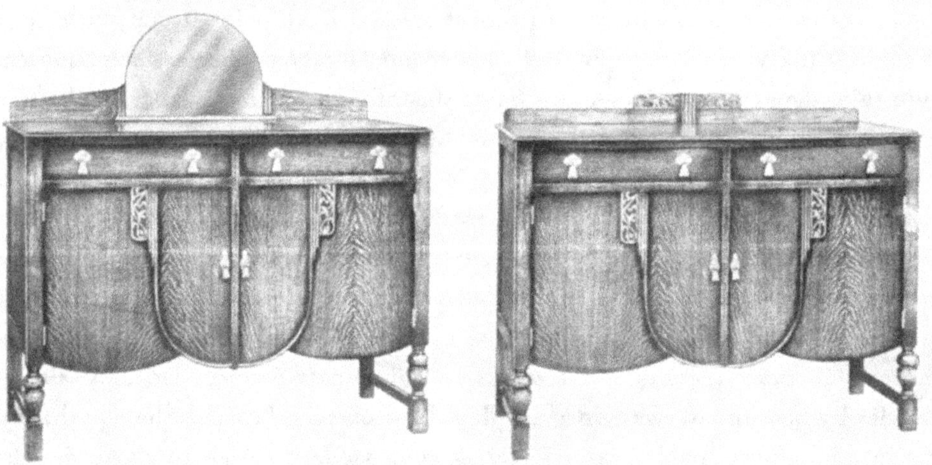

XXIV. *Best selling type of modernistic carpet. Taken from an illustration in Mr C. E. C. Tattershall's book "A History of British Carpets".*
Best selling pieces of furniture (by courtesy of the Co-operative Wholesale Society).

can less afford to take risks than a manufacturer or an independent shopkeeper. Moreover, the class from which most travellers, buyers, and smaller shopkeepers come is supposed to be one of the most reactionary classes in the country. The houses in which they live and the prejudices with which they grow up make them less accessible to modern art than the professional as well as the working classes.

In these circumstances the manufacturer, supposing he is not an exponent of the same good-old-days mentality, ought to try to find out whether the information which he receives from the sales department is a true interpretation of the wishes of the public. He can do that easily in the few cases where a firm has shops of its own. Otherwise he can attempt to check his travellers' reports by sometimes travelling himself, or by going to shops and enquiring privately, or by showing new models to as many friends and employees as possible, or by organized experiments of the type carried out by Messrs Cadbury and Messrs Foley China, or by trade questionnaires such as I mentioned in connection with radio firms, or by other questionnaires on a large scale. I quoted the results of my own experimental questionnaire, which sufficed to show that it would be worth some money and trouble to try something of the sort in a more expert way. Much too little has been done by British producers to obtain a more objective view of the popular taste. The results of any such enquiry would no doubt be surprising. We saw that in some trades, such as door furniture and cheap electric fittings, the wishes of the public hardly count. Almost all business is transacted with builders' merchants. In buying or renting a cheap house, it is only in exceptional cases that the client will express his view of the door handles, or of the wall brackets and pendants, which he sees in the rooms. Here then the popular taste—and, as I said, a distinctly hideous taste—is not the taste of the public but that of the builder and his merchant. One may say that this probably comes to the same, but it ought not to be forgotten that the same builders have recently shown a predilection for panel fires, i.e. a type of fire greatly improved in design, and for plain distempered walls (notwithstanding the fact that when they feel it their duty to be artistic they go in for excesses of appliqué wallpaper). The same type of wholesale merchant in the jewellery trade, though he insists on conventionally and thoughtlessly designed brooches and bracelets, will refuse wrist-watches which are not simple and plain. To explain this apparent contradiction I can only refer to previous remarks on objects for use and objects for adornment.

The same strange contradiction is to be found in the purchases of the public. We saw, to take the most alarming case, that the best selling textiles are fairly simple weaves and fairly well, or at least inoffensively, designed printed cretonnes on the one side, and horrid modernistic carpets on the other. How can it be that the same class goes in for both? The reason here, and that is why the question appears of considerable importance, lies undoubtedly less with the consumer than with the producer. In printed fabrics a genuinely English tradition was so strong as to keep modernism out. In carpets the corresponding Oriental tradition was not inborn but imported from afar, and therefore too weak to resist the attack of post-war jazz. Manufacturers accepted the fashion, shops were swamped, and the public followed without hesitation. There can be little doubt that things would have gone differently had manufacturers held aloof from "modernism". The fact that sales of plain carpets have gone up at the same rate as sales of modernistic patterns shows that there is a demand for something simple and better.

In some other trades the decisive importance of the manufacturer may emerge even more clearly. I referred to streamlined cars in that connection, also to steel furniture and to modern banded, as compared with modern floral, pottery.* In each of these cases producers have deliberately pushed a certain line, pushed it consistently and over a sufficiently long time, with the result that it became the fashion and the quantities of sales expected could be secured. I say over a sufficiently long time, and that is an essential point. We heard that Messrs Ferranti had to wait some years before their new type of fires became popular; in banded pottery the time-lag between the start and the universal acceptance was about three years; in streamlining, between Sir Denistoun Burney's initiative and the conquest of the English market, about four years; in English tubular furniture about four years again. This shows how absurd and unpardonable a practice it is in some stores to give a new article one month's or three months' trial on the counters and then to drop it for ever if sales have not reached a certain level. To call this investigating the popular taste sounds like satire.

Considering all these conflicting facts and arguments, the only statement on the public taste which can be made is that "the public does not know itself what it wants",[1] whereas most distributors and manufacturers know pretty well what they want. A certain responsibility of the consumer cannot be denied, but it is negligible as compared with the producer's and the retailer's. If we need another case in point, it is once more the glass department at Woolworth's. Do the customers, because the goods are too well designed, express their abhorrence by boycotting them? Not in the least; they buy them exactly as they buy the badly designed articles under the same roof. Perhaps after a few more years of propaganda, perhaps after a generation, they would no longer relapse into certain atrocities. But under present conditions, their taste is a blank. They take what they are consistently offered. Ruskin has preached that the business of the manufacturer is to form the market as well as to supply it. He might have added that whether the manufacturer wants it or not he is forced to form the market, for better or for worse.

However, a later section will be devoted to the duties of the producer. In our present connection, it would not be fair to deny some obstacles to perfect design in the public itself. A certain admiration for elaborate craftsmanship of the same type as that found in the sixteenth-century crucifixions with fifty-two persons present, all carved in a cherry-stone; or of modern Canterbury Cathedrals made by enthusiastic amateurs from cork, or lard, or sugar, is a natural outcome of a simple, unsophisticated mind. So is a certain admiration for "Kitsch". On the other hand, perfection of design cannot be welcome to most human beings, because it is too exacting to live up to. To see Raphael's Sistine Madonna, Michel Angelo's ceiling, Rembrandt's Christ at Emmaus, Grünewald's Isenheim Altar once a year can be a sublime experience, but one would not like to live in the same room with any of them. However, allowing for human frailty is still no excuse for jagged door handles and vulgarly decorated cups and saucers.

It is agreed that the public could not buy these unless they were put on the market. But if people disliked them as thoroughly as one would wish them to, they would not buy them even though they were on the market. The fault of

[1] J. Gloag, *Design in Modern Life*, 1933.

the public in the matter is that very few people take as much trouble in buying a lamp or a vase as they do in buying a hat or a tie. Not that there are not vast numbers of people who really enjoy the atrocities which shopkeepers display. They are accomplices to the crime, no doubt. But while there is no justification for the educated and wealthy managing director of a large firm producing rubbish, there are several excuses for the humble consumer, excuses derived from the social conditions in which he is compelled to live.

THE SOCIAL PROBLEM

No exact figures are available as to what proportion of the English population grows up in slums and semi-slums. Official reports tend to be too lenient. An independent investigation carried out in Birmingham in 1933–4 showed that 28,000 houses should be cleared at once and 18,000 back-to-back houses as soon as possible.[1] The word slum has become a common term for unbearable housing conditions all over the world; rickets, the disease mainly bred in slums, is called the English Disease in Germany. Not that overcrowding and its consequences are confined to Britain in any way. Things are equally bad in other countries, worse in some, better in a few where great building activity was possible immediately after the war. A recent questionnaire filled in by 900,000 schoolchildren in a country which is supposed to be exceptionally clean has brought out that 39·4 per cent. of the children use no toothbrush and 18·2 per cent. a family toothbrush.[2] No wonder that in such conditions no sense of beauty and hardly a sense of tidiness can be developed. And no wonder either that the natural longing for beauty, inborn in almost everybody, slakes itself on the cinema, on showy furnishings and "kitschy" pictures in the house and on the portraits of film stars—heroes and heroines of dreams of wish-fulfilment—in the workshops.

Design in daily life is not a detached question, this cannot be said too often. No conclusive improvements can be made in design unless social improvements are achieved before and alongside them. A sweeping change of social conditions,

[1] Cp. *Birmingham's Black Spots*, published by M. W. McIver and L. S. Florence; privately printed by the Hereford Times, Ltd.

[2] Health Service of the Museum of Hygiene, Dresden, 1934.

such as the establishment of some kind of State Socialism, might lead to a sweeping change in the appearance of industrial products, although it seems impossible to generalize on the aesthetic consequences of such a change. In Germany the post-war Labour Government fostered the modern style, whereas National-Socialism tends towards the protection of handicraft as opposed to industrial design. In Italy modern architecture and modern design enjoy the special furtherance of the Government as being an expression of Fascism. In Russia the same style was, after some years of State approval, given up as an outcome of latter-day bourgeoisie, and a decidedly conventional and therefore probably more popular style launched. Whereas the reaction of art to changes within our civilization thus remains indeterminate, the consequence of a breakdown of the whole of our civilization and the beginning of a new circle is evident. The "Merovingian Age" which would necessarily be the first phase to follow would at one blow get us back healthy though primitive conditions of craftsmanship, the unity of design and production, and Morris's "joy in making" which is definitely lost as long as present conditions last.[1] It is a matter of personal outlook and taste whether one regards the price to be paid for a true revival of art and craft as too high. In any case, it is outside the scope of this book to discuss standards of design as they will prevail in a new era which will no longer be part of what England calls "Western Civilization". My task here is to describe the present situation and to suggest improvements that are practicable in the twentieth century.

Among these, slum-clearance and re-housing are of the utmost importance. Recognized as such in England earlier than anywhere else and courageously faced by a few pioneers such as Lord Ashley, the founders of the Peabody and Guinness trusts and the creators of the first working-class estates (Sir Titus Salt) and the first garden suburbs (Messrs Cadbury Bros. at Bournville, Messrs Lever Bros. at Port Sunlight), they are still as imperative as ever. The contrast between "the fields where beasts live and the streets where men

[1] "How often it consoles me to think of barbarism once more flooding the world" (W. Morris, in J. W. Mackail, *The Life of William Morris*, London, 1899, vol. ii, p. 144). "Perhaps the Gods are preparing troubles and terror for the world . . . again, that it may once again become beautiful and dramatic withal" (*idem*, vol. i, p. 305).

live"[1] has by no means been removed. Parallel with re-housing, there ought to run an equally energetic campaign for cleaner streets, cleaner and better planned suburbs and above all for cleaner children and cleaner parents, and for healthier and more modern school buildings.

New school buildings will remain insufficient, however, as long as the contrast between the expensive public school and the Council elementary and secondary school is not only a contrast of building and surroundings, but also of the character of education administered.[2] And just as the public school is—in spite of the growing number of scholarships—a privilege of a few, so is the possession and the use of so much of the English countryside and also of so many of the open spaces in inner London. No wonder that the "something-that-is-different" attitude thrives in this atmosphere.

So it remains doubtful how much of the desire for better design which we have noticed, and which is evidently growing, is based on a genuine appreciation of beauty, and how much on a craving for visible social distinction. Often it is bewildering to see that the same wealthy industrialist whose house is furnished in a pleasant taste, works in a dingy office which you reach by driving or walking through filthy streets, passing some black sheds with blind and broken windows and climbing up a dark, narrow and dangerously steep staircase. To explain this strange contradiction, it is necessary to refer to the deep-lying tradition which Puritanism has become in England. Work is our duty, but duty is something stern and forbidding that one ought not to embellish. There is something of the same attitude in the social position of English painting. The glory of English eighteenth-century art lies in its portraits, i.e. the very genre in art which can be excused by regarding it as not a mere matter of delight.[3] Again, the beauty of pre-Raphaelite pictures is so completely detached from everything that mattered at the moment when they were created that their lasting success can be accounted for only as a means of escaping real life. Art ought to remain a permissible relaxation, and life a vale of tears. It is extremely characteristic

[1] W. Morris, *The Lesser Arts*, Collected Works, vol. xxii, p. 25.

[2] Lethaby on public school education (*Form in Civilization*, p. 133, 1922): "Its inward unrealized function was to provide a myth of superiority to those who could pay."

[3] To understand the importance of landscape in England, other national qualities have to be taken into consideration. This, however, is outside the orbit of our problem.

that it was the big industrial towns such as Birmingham and Manchester that bought the best collections of pre-Raphaelite works. But do we want this attitude to go on for ever? Must railway stations remind us of purgatory? The new underground stations in London prove that England is well able to build them as attractively as any Continental country. Why should so many shops shrink from an adequate display in their windows? Surely there is no reason here for applying the Puritan attitude which distrusts beauty as something questionable and unreliable.

It has appeared to me advisable to call attention to many changes that are urgently needed, because occasional improvements of design cannot be of more than ephemeral value so long as conditions of life remain as they are. But it would be a grave error to leave industrial art alone for the time being in order to devote all available energies to other fields. The battle has to be fought on all fronts. Not one of the subjects is less essential, not one can be neglected, neither slum-clearance nor the renovation of school buildings, neither the levelling up of class contrasts nor the raising of standards of design. It is only with this last aspect of the problem that the remaining pages of this book are concerned.

SUGGESTIONS FOR IMPROVEMENTS: I. ART EDUCATION

The soundness of a building depends on its foundations. If serious flaws become visible it is advisable first to look at the foundations. Now as to the artistic standards of designers, manufacturers and public, the first foundations are laid while they are children, go to school, and have their lives centred in schoolrooms. The Pick Council realized the vital importance of this and devoted its first pamphlet to educational questions. The report describes the teaching of art (and handwork) in infant, junior, senior, central, secondary, and public schools, the supply of specialized teachers, the training available for them, and the equipment of schools. I have been allowed by H.M. Stationery Office to reprint the results of this careful and comprehensive research *in extenso*:

"In senior, central and secondary schools Art should have the same measure of attention as is given to languages, science or mathematics. In junior or infant schools we have found a consensus of opinion among witnesses in favour of approximately one-fifth of the total school time for art. We think that serious consideration should be given to the matter from this point of view. Periods devoted to art should be fairly long and the size of the class limited. Facilities should be given for voluntary work outside school hours.

"Wood and metal work provide a means of learning to appreciate good design, and should be made to constitute a lesson in design. An end should be put to the present divorce between the teaching of wood and metal work on the one hand and the remaining subjects in the Art curriculum on the other, by the proper association of these subjects in the training of the teachers and the arrangements for inspection. Art should influence the teaching of handwriting and the illustration of lessons in history, geography, literature, nature study and domestic science.

"In order to counteract the present tendency towards lop-sided secondary and public school education, Art should be given a more important place as a subject in the School Certificate examination—possibly as a 'major' subject for pupils most gifted artistically and as a subsidiary subject for other pupils. Some alternative to a paper examination might be considered.

"Art should be a compulsory subject in the training colleges. Short courses for existing teachers whose training did not provide adequately for instruction in the teaching of art should be further developed. The 'Art and Handicraft' section of the proposed revised edition of the Board of Education *Suggestions for Teachers* should be published without delay.

"The employment of specialist teachers of Art should be encouraged in senior and central schools as well as in secondary schools. Teachers of wood and metalwork should be able to teach their pupils the application of good design. Opportunity should be given to specialist teachers to teach general subjects.

"The supply of teachers of wood and metalwork from Loughborough, Shoreditch and other training colleges undertaking specialized courses should be increased. The provision of third-year courses at these training colleges in collaboration with the Art Schools should be extended. Further provision should be made for courses for existing teachers and for classes for teachers at the local Schools of Art.

"The School Certificate or Matriculation requirement for entry into a training college should be relaxed for the bona fide entrants from industry. The training colleges should extend the range of handicraft now taught. The City and Guilds Institute should consider the desirability of adding to their examination for teachers of wood and metalwork tests in design and the cultural aspects of handicraft (the certificate granted to the candidate being endorsed accordingly) and of imposing an age limit on candidates.

"The training colleges should provide a three-year course for all specialist teachers of Art, and the third year might with advantage be worked in conjunction with a local Art School. The Art Schools should also be developed as a training ground for specialist teachers of art and craft for senior and central elementary schools as well as for secondary schools.

"Local Education authorities should give serious consideration to the improvement of the children's surroundings, and particular attention should be paid to the re-decoration of old school buildings which it is not possible to replace. There should be separate, commodious and suitably furnished Art rooms, adjacent to the handicraft rooms, in the schools wherever possible. Authorities should regard it as a matter of urgency to provide elementary schools with collections of well-designed common objects for the lessons and suitable facilities should be available for the display of such objects and of pictures."

All this is excellent and hardly anything can be added. At present, it seems to me, the teaching of art in certain elementary schools is more advanced and more alive than in the majority of public schools. While the vast improvements made here within the last few years appear to be connected with Continental theories and experiments (Professor Cizek and his classes in Vienna, to mention only the one case which, owing to exhibitions in England, has become especially well known),[1] in the teaching of handicraft this country has for a long time been ahead of the Continent. In reading and examining the suggestions contained in the report of the Pick Council, one should not forget that only distinctly modern-minded schools on the Continent give as much serious attention to the crafts as is usual everywhere in England. As for a live appreciation of artistic values in general too little is still done both on the Continent and in England. Secondary

[1] Cf. W. Viola, *Child Art and Franz Cizek* (Austrian Junior Red Cross, 1936).

schools should try more than they do as a rule to open up for the pupils an approach to the aesthetic qualities of works of art and architecture. Those who have been introduced to the beauty of line and composition found in a Gothic figure or a Holbein drawing or (this is indispensable) in an outstanding piece of contemporary architecture will not be satisfied in later life with the vulgarities that they are offered in the average shop and store.

As things are to-day, the future skilled worker and designer enters his trade provided only with what the elementary school has given him. In some cases he may pass three years in the junior art department of an art school between form-room and bench. There are great possibilities in the system of full-time junior courses, although they have necessarily to consider rather the worker's than the designer's outlook. For the training of taste and appreciation they could be invaluable, if handled in the right spirit by the right men. Their importance is all the more evident in England, because an essential and well-known difference between the English system and that in most Continental countries is that England has no compulsory continuation schools. It depends on the personal goodwill of the employer whether a youth gets an afternoon or a day off for improving his taste and his knowledge of his trade in general, or whether, if he feels a desire to do so, he has to go to evening classes after the end of his day's work. The Board of Education emphasized the importance of continuation schools as early as 1910.[1] Another report of the Board, in 1932, discussed the excellent effects of the French compulsory *Taxe d'Apprentissage* which is spent on art education.[2] It is beyond doubt that some arrangement of this kind ought to be introduced into this country. If manufacturers are not of themselves sufficiently social-minded to see that here is a duty of theirs, then a new law should help. The argument: "What is the good of this to us? We want hands"[3] is criminal, doubly criminal to-day, since the problem of the use of leisure has become so urgent.

I visualize art schools and technical schools in England, provided with compulsory afternoon courses for apprentices in all trades, and freed from much of

[1] H. A. Clay, Board of Education, Educational Pamphlets No. 18.

[2] A. Abbott, *loc. cit.* No. 91.

[3] C. R. Ashbee, *Craftsmanship in Competitive Industry*, p. 188, 1908.

the cumbersome work now going on in the evenings. But I do not by any means visualize the organization of art schools in this country entirely remodelled on Continental lines. As I said in Part I, most European countries have separate continuation schools, trade schools, schools of arts and crafts, and art academies. In England as a rule all these purposes are served by the same school in the same building. I firmly believe that this system is bound to be beneficial, and I should even like to see it extended so that a closer co-operation between art schools and technical schools would everywhere result. It cannot be anything but extremely advantageous to have the training of artists, architects, craftsmen, engineers, and skilled industrial workers under one roof. This should protect schools from becoming "arty" as well as "tradey".

A few more words must be said about the training of designers, which is unquestionably one of the principal objectives of art schools in our century, an objective of much more weight than the training of the independent craftsman, the painter, or the sculptor. The independent artist-craftsman has too little scope to-day to justify the provisions for his training to be more than a side-line in a modern art school. And the same, I am afraid, is true of the painter and sculptor. One may not like this fact, but there it is. Demand for pictures and sculpture is small, much smaller than it was before the industrial revolution which, by the invention of photography and colour reproduction, has so greatly affected the social conditions of art. Before a complete upheaval of our civilization destroys all modern machinery, including the inventions just mentioned, this situation will not change, and the artist, as well as those responsible for art education, must face it. It is wrong, sociologically and morally wrong, to base the organization of art schools on provision for future painters and sculptors. It helps to create an art proletariat without really furthering the genius, who is nowadays the only artist that matters, and who cannot be materially furthered by any organized training. This is the way things are at present, at least in England. The situation is somewhat different in the new "totalitarian" States, where the government must have a strong interest in encouraging as many artists as they need for creating works of a certain character and in conformity with a certain programme. This naturally gives a new scope to schools of fine art.

For England, the most adequate organization of art schools to-day would be something like this: There ought to be classes for design, and classes for art. In the design classes the execution of articles for all crafts and trades should be a matter of course. Also it would be highly desirable that such articles should be sold in competition with those produced in the ordinary commercial way. The one strong objection against this practice, which is in use in a few Continental art schools, is that a school on account of public grants can work more cheaply than a private manufacturer and that its competition in the open market would therefore be unfair. So explicit is the feeling about this in England—and the same opposition has also arisen time and again on the Continent—that I noticed a certain fear amongst principals of art schools that they might be deprived of all private support directly they took any steps towards such an aim.

However, I think art schools themselves could invalidate this argument by correctly calculating and rigidly adhering to the general price level. This they should do not only for the sake of their market but also for their students' sake; for commercial calculation is one of the matters which the designer ought to be able to learn in the school. I do not underestimate the practical difficulties which the introduction of any scheme of school production for sale would have to overcome. But I also do not believe that it would commercialize schools and induce them to give up their standards. On the contrary, I think it would react favourably not only on the spirit of the school but also on the prevailing taste in commercially produced industrial art. The schools would have the advantage of being faced with real life and could no longer be mistaken about the conditions under which their students will have to work after the end of their studies. School products would be constantly tested as to their practical qualities and public appeal. Industrial art in general would on the other hand gain immensely by the competition of goods designed to standards as exacting as only an art school would feel obliged to hold up. Interference of sales managers and travellers would be reduced to a minimum, and the artistic quality of objects thus conceived could not fail to act as a stimulus to private enterprise, as has in fact been the case in several Central European countries since the war.

So much for design classes. With regard to art classes, I do not think it would be advisable to abolish them entirely as some extremists demand. A better method appears to be that which was developed immediately after the war by the Bauhaus and some other German art schools, namely to let every future artist first pass through a complete craft or trade training in the school.[1] If that deters superficial talents from continuing, all the better. And all the better also if it induces as many students as possible who entered the school with the idea of becoming painters, to change their minds and devote their gifts to design. Such a method will be passionately attacked by the few real artists worth that name—wrongly, for they ought to realize that what they want they can never get in classes, but only by direct intercourse with great men. It will be for the authorities concerned with art education to find such men as teachers for art schools. And after all, as far as England goes, there are always the Slade School and other independent schools as refuges of free art.

As to craft classes, they could be the same for students who want to become artist-craftsmen and designers, though the emphasis ought to be laid on the needs of the designer. That means that craft classes must be regarded as laboratories for experimenting with materials and basic processes.[2] They can no longer be an aim in themselves, as Morris and C. R. Ashbee—in many ways the most consistent of his followers—had preached,[3] but they are of vital and indispensable advantage in conveying to the designer a live understanding of the natural will of a material which does not reveal itself except to those who shape it with their own hands. But craft classes are only one part of the training which designers require and should be, from the beginning, supplemented by courses in basic composition. Elementary architecture is also of the utmost importance to any designer, and it is greatly to be regretted that in England the teaching of architecture is so detached from that of the other arts and crafts. In architectural classes the student is much more likely than in craft classes to get some insight into modern materials, modern processes and commercial calculation, all of which are essential to the future designer.

This leads up to the question of equipment. No doubt a student cannot be expected to develop a sense of machine requirements as is needed for the

1 Cf. W. Gropius, in *The Year Book of Education*, 1936.

2 Cf. W. Gropius, *loc. cit.* and F. L. Wright in his answer to the *Studio* questionnaire, *loc. cit.*

3 C. R. Ashbee, *Should we stop teaching Art?*, London, 1911, especially pp. 99ff.

creation of standard shapes, if he cannot get any information about the kind of machinery which is going to carry out his ideas. As for English art schools, it has already been stated earlier how inadequate is their equipment. I referred to the Kidderminster School not using a Jacquard loom, and to the Stourbridge School not being able to cut or fire glass. I also suggested that in some cases co-operation with existing technical schools or colleges could to a certain extent make up for this deficiency.

To attain any more changes it would be necessary to secure the active help of State and Municipal authorities, of Trade Associations and individual manufacturers. If the manufacturers in most towns are still distrustful and unwilling to take an interest in art education, the fault is but partly theirs. In a previous section I have pointed out that in many schools craft classes are still distinctly "arty-crafty" and design classes completely absent. Where the new system and the right men are at work, conditions will rapidly improve. Leicester and certain classes in Birmingham could serve as proofs for the purpose of convincing industrialists.

Once a school had secured better equipment it would be infinitely more successful not only in its design course but also in the afternoon trade classes. For them modern machinery appears particularly important, because the aim of these classes, whether of continuation school character or attended by skilled workers in industries, is, as I said before, to give an all-round insight into their industry to all those youths and men whose daily routine work deprives them of any comprehensive understanding of what they are actually engaged in.

As to other activities of English art schools, courses to be given to future teachers should also include basic composition and architecture besides crafts and drawing. It would be advisable to add lectures and "seminars" on the appreciation of art. Incidentally, it is scarcely necessary to add that—as emphasized by the Pick Council—the separation between the teachers of art and of handwork should be completely abolished.

Moreover, some general courses ought to exist everywhere for manufacturers, buyers, commercial travellers and salesmen. They should be held in the evenings, or better in the afternoons, and it should be made the duty of manufacturers and distributors to send their employees.

After these general remarks, some suggestions may be added as to the spheres of activity to be assigned to the various types of art school. I prefer to leave private schools entirely out of account, although in this country of private enterprise a Bauhaus re-created at the present moment as a private school could have the most stimulating effect. As to municipal art schools, these should be developed further with a more pronounced bias towards local industries. Drawing and modelling should in most smaller towns be regarded as subsidiary subjects. The training of painters and sculptors ought to be limited to a few schools in the centres of cultural life. In towns with one main industry, the school should come as near to the Monotechnic type as feasible. By such specialization expenses would be reduced in several respects, and thus the funds available for acquiring adequate plant would increase. In large cities with several important industries several Monotechnics (of the type of the Birmingham School for Jewellers and the London Shoreditch School) might be run in connection with branch schools under a Central School.

In London, a closer collaboration between the Central School and the other County Council art schools seems desirable. The Central School deserves to be developed into a centre of design education not confined to only a few trades. Here more than anywhere else, modern machinery ought to be purchased on an ample scale, so that a complete training could be given in modern methods. Different classes should be attended by skilled workers and by future designers. Since the demand for highly qualified designers is limited (and would become even more limited, if more firms decide to use first-rate men on a part-time basis), selection ought to be as severe as possible. Design students would either come from secondary schools into the design course, or else be drawn from the trade classes.

As to the future of the Royal College of Art, I am somewhat at variance with the Memorandum of the Design and Industries Association, mentioned earlier. The Memorandum suggests developing the Royal College into a "National University of Industrial Design", omitting, however, certain industries the training for which could be better organized in the towns where these industries are localized. The system which I visualize would provide for completely equipped art schools of a Monotechnic or Polytechnic character in all such towns, including London, and leave to the Royal College the most advanced education only.*

Students entering would already have a knowledge of the technical methods in at least one industry, and also a thorough craft training. The task of the Royal College would be to give them a wider outlook, a more adventurous spirit, in short those qualities that would enable them to secure the highest and most responsible posts in firms concerned with artistic design. It is evident that to achieve this task a perfect equipment with modern machinery for all industries is by no means indispensable. For the needs of an "Academy of Design", as the Royal College should be, it would be sufficient to keep up a regular co-operation with the London Central School, some other specialized London schools, and with a small number of enlightened manufacturers in whose works the creative young designer could experiment on the practicability of his designs in particular materials and at a particular price. I do not see any reason *a priori* why such co-operation should be unattainable, provided the right men are in charge of the schools and the classes.

Thus we are once more faced with the fact that improvements depend more on personalities than on regulations. That the Board of Education also realizes this is shown by its policy during the last few years. But how about local Education Committees, especially in smaller towns? One means of getting teachers of a higher status would be to pay them better salaries, to allow them to continue their former activities as designers or craftsmen, and to give them tempting studios of their own in the schools. This above all—a practice quite common on the Continent—would make a great difference and attract many a man who now shrinks from leaving London to bury himself in a gloomy Victorian school building.

II. EXHIBITIONS—PRIZES—REGISTRATION OF DESIGN

The importance of exhibitions as a means of propaganda has been discussed before. To examine this question was the principal aim of the Gorell Committee. Whereas the Committee pleads for a permanent exhibition in London, the Industrial Art Committee of the Federation of British Industries considered annual exhibitions more useful. I believe that both are right. There should be in London a proper Museum of Modern Design, either as an independent body or as a part of the Victoria and Albert Museum, housed in a modern building and arranged in the most up-to-date way. It should collect objects of industrial art mainly and include pieces of individual craftsmanship only so far as they are of experimental value to industry. It should possess a selection of goods large enough to allow for frequent changes of display.

But beyond this there should be a big annual exhibition enjoying if possible all the public interest and all the publicity attached to the Ideal Homes, the Motor or similar exhibitions. And since provincial towns are even more in need of good exhibitions to counteract the usual type of local trade show, it would be of primary importance to send round the country travelling sections from the annual exhibitions, as was done at the end of the Dorland Hall Exhibition in 1933, and as the Pick Council is going to do on a larger scale.

As to the organization of such exhibitions, which are meant to represent English Industrial Art at its best, I said in Part I that it seems fundamentally wrong to entrust this work to vast committees, in an attempt to avoid any personal standards. On the contrary, a muddled impression is bound to result unless the risk is taken of appointing one conscientious and distinguished expert belonging to the war or post-war generation (or, if that seems too dictatorial, a "Directoire" of three), for the whole of the selection. In this respect, the lesson taught by the Burlington House Exhibition last year was sufficiently clear.[1] The task of the organizer of an exhibition such as I visualize would be to act on his own standards, and to keep them rigidly and uncompromisingly in choosing from existing commercial English production that which deserves to be shown. Whether an article is new or is already on the market ought not to be of any importance. But whether it is machine-produced or not ought to be seriously though not narrow-mindedly considered. And whether it is available in ordinary shops at a reasonable price ought to be one of the main conditions of selection. Naturally even the standards of the most careful and conscientious expert will be personal, and not exactly identical with those of other equally reliable experts. That one should not mind; on the contrary, it would make discussion more forcible and the general impression more lively. It would also lead to certain differences of character between the products shown in different years. However, one quality ought to be indispensable for all objects to be exhibited. They must

[1] "Apparently no one took it on himself to direct what was to be done at the exhibition" (Mr Frank Pick, in *Journal of the Royal Institute of British Architects*, p. 588, 23 March 1935).

not be on the lines of period styles. It seems to me a great mistake to argue that old or new in such an exhibition is immaterial, because new things will soon be old and therefore neither better nor worse. Worse perhaps not, but just older, and that means representing the style of another generation, another period. The aesthetic qualities of a work of art certainly remain unchangeable, but its most essential quality changes—its vitality.[1] We do not want to exhibit objects lacking in vitality, objects that are not frankly of our age.

Another criterion for admission to such an exhibition, and one which has been much discussed of late, is the national origin of the objects shown. I consider it a sound principle to reserve a national exhibition for national products. The character of a nation is undeniably one of the paramount causes of its artistic idiom. People without much experience often call the style of modern architecture international. But those who have seen more and thought more about it, will find it easy to recognize in modern Paris houses French *charme*, wit, grace, and at the same time French delight in abstraction and pure form, and in Gropius's houses German severity, sternness, thoroughness and social fervour. They will find it equally easy to distinguish between the neo-Classic expression in modern British (Sir John Burnet, Lloyds Bank, 1928–30) and modern Italian (Marcello Piacentini, Bergamo, Littoria) architecture. There is, however, one doubt as to the wisdom of applying the principle of nationalism to exhibitions of industrial art. The Modern Movement, as I have tried to prove, not by any means un-English in itself, is still too new in this country not to need contact with foreign work, which, owing to our modern means of communication, it could not be prevented from obtaining, even if this seemed desirable. It is far better to make the best foreign production available to English designers and manufacturers, in order to sharpen their discrimination. In these circumstances, the best suggestion might be to reserve the large annual exhibitions to national products, and to stage in addition smaller shows of foreign goods (the products perhaps of only one or two trades at a time) to be held in London and the provinces.[2]

[1] "The essence of all the old arts was in their vitality, their response to the natural conditions and the psychology of their times" (Lethaby, *Form in Civilization*, p. 12, 1922).

[2] The Pick Council suggests such exhibitions but does not want them to be open to the public. From my point of view there is no sufficient reason for shutting out an interested public. On the contrary, I should be inclined to suggest the Museum of Modern Design as the most suitable place.

In connection with the National Exhibitions of Industrial Art, governmental prizes should be given to outstanding objects, and purchases should be made by museums in London and the provinces. For permanent collections of contemporary industrial art ought to be started in all those smaller towns which are specializing in one industry. Prizes are also of great importance at art school exhibitions.

Another necessary initiative to be taken by the State, in order to encourage better design, is a reform of the laws governing the registration of designs. Everywhere I heard complaints that it is hardly worth registering a pattern or a shape, because the slightest alteration would entitle a competitor to crib the new model with impunity. Here the aid of Industrial Psychology might once more be useful. It appears possible to introduce laboratory tests as to whether a supposed copy is mistaken for the original or not. A practicable method might be to show samples or slides of a series of articles, say twenty decorated plates one after another. Sixteen should be different, one of the sixteen being the design to be tested. Mixed in among them, the alleged copy would be shown. In order to make the experiment less simple and thereby give the result a higher degree of validity, three designs ought to be repeated in the course of the experiment. The test would be, whether the majority of the people asked regarded the alleged copy as a fourth repetition or as a different design. Perhaps there are objections to such tests which are unknown to me. Costs, at any rate, would not be too high.

III. OTHER PUBLIC INSTITUTIONS—PRIVATE INSTITUTIONS —PUBLICATIONS

Great improvements could be achieved by the Design and Industries Association, if it were possible to enlarge its membership, to renew its organization and to strengthen its financial power. One means of improving its position would be the introduction of an entrance jury which would see to it that only men of real artistic conscience could become members. That would mean a quantitative loss for the moment, but one ought not to underestimate the amount of ambition and competition which might, by this device, be roused. It should become a desirable honour to belong to the D.I.A. and be allowed to use its initials, as was the case with regard to the German Werkbund before the war. Another way to obtain greater publicity for the Association would be to rent a stand at the British Industries Fair and show there some of the year's work of the members in a good architectural setting. Such a room would have the same effect as the famous "Grassi-Museum" section on the Leipzig Fair, a section selected twice a year by a jury of independent art experts. It would be invaluable for those buyers who are interested in good design and who now never know beforehand where they will find well-designed articles.

Trade Associations could also help a good deal, if they wanted to. So far the attitude of most of them has not been very encouraging. This year's Coronation mug, the result of a competition held by the British Pottery Manufacturers' Federation can hardly be called an example of commendable modern design. And what will the other Coronation souvenirs be like? One recent organization that was founded by industrial interests seems to do very promising work: the British Colour Council, formed to help British fashion trades but developing beyond this aim and active in improving the colour of industrial products.

Newspapers could also help considerably, and often do. The B.B.C. is a very strong power, far stronger than art magazines, though their persistent propaganda certainly has a more far-reaching effect than one would expect from the mere number of subscribers. Another great force for better or worse are the railway, bus and tram companies. If they decide to work for a decent taste in their cars and carriages (as indeed the London Underground is doing, thanks to Mr Frank Pick), and for a decent taste in their tickets also, they cannot fail to influence the taste of the public in the end. Postage stamps are also of importance. Until 1935, they had not even freed themselves from the nineteenth century's attitude of vague period imitation, a remarkable document of English official belief in once established forms. Then came a first very cautious improvement in the Jubilee stamps, and a much more energetic step in Edward VIII new stamps. Although these were not designed by any competent artist, the arrangement of photographed head and lettering was clear and on the whole quite satisfactory as well befitted a king so sincerely interested in modern industrial and decorative art. Still, they were by no means perfect either; and it is to be hoped that for this year's new stamps the objections rightly raised and widely discussed will be duly considered.

THE DUTIES OF THE CONSUMER

I only need here to recapitulate what has been said before. If consumers in general would firmly refuse to buy an ugly thing simply because it is the only one of its kind in a shop, if they would make up their minds "not to have anything in their houses that they do not know to be useful, or believe to be beautiful,"[1] and if they would select a kettle or a clock as carefully as an evening dress, the worst would be avoided.

Secondly, if wealthy and well-educated consumers could be brought to understand that buying reproduction work means fleeing from contemporary life, and buying genuine period pieces means destroying the market for good modern pieces and thereby reducing their poorer fellow-men to badly imitated period or vulgarly designed modernistic things, and if they would begin to feel it as their duty to set an example in taste not only in dress but also in domestic design, another way to improvement would be opened.

Thirdly, a word on the great possibilities of support for the Modern Movement in connection with the royal family. Several times manufacturers told me that the purchases made by members of the royal family at the British Industries Fair had set a particular fashion or decided between two possible lines. If here a decidedly modern and well-advised buying policy were introduced the reaction on consumers and producers would be immediate and no doubt beneficial.

THE DUTIES OF THE MANUFACTURER AND THE RETAILER

In order to attain improvements of design without losing sales, the manufacturer must, as I said earlier, try to obtain an exact knowledge of the popular demand without relying on the retailer, buyer and traveller. I indicated before that this can be done by statistics, by showing new articles to representatives of various classes and collecting opinions, and by questionnaires.[2] All such means would enable the industrialist to follow the line of least resistance in the improvements that he wants to introduce.

[1] W. Morris, *The Beauty of Life*, 1880, Collected Works, vol. xxii, p. 76.

[2] I hope that in the near future it will be possible with the aid of an English university to start on a questionnaire investigation more comprehensive and more reliable than any which so far have been undertaken in this country.

After having secured information as to the Where? the manufacturer must concern himself with the How? Here I have to refer to previous considerations about the different methods used for obtaining designs and the great advantages that lie in the part-time employment of architects and modern commercial artists. On the other hand, even those manufacturers who want to do their best claim that they cannot limit their production to well-designed articles, because—at least to-day—these would not suffice to keep their business going. This is true, I am afraid, and one has to think of acceptable ways for combining "artistic" and "commercial" production. Several instances were given in earlier pages. The simplest is the case of the firm manufacturing electric fittings, the owner of which admitted that a large part of his wares was at variance with his own standards, but pointed out that only by making money out of ordinary and bad design could he afford to try and sell interesting modern models. I found a similar though less marked contrast between the commercial production of one of England's best-known manufacturers of furnishing textiles and their affiliated firm of highest quality fabrics, and also between the ordinary suites in a high-class furniture works and their one or two admirably designed ranges of modern articles. Nearer to the perfect compromise (in present circumstances) appears to be the system adopted by one of England's best furniture workshops and by one of the factories making steel furniture. In the one case the production of extremely well-designed wireless cabinets on a mass-production scale, in the other the production of bus-seats, serves as a "bread-and-butter line" to give a margin of profit safely to run the other side of their business.

In this connection it may be useful to mention one distributor at least who works on similar lines. I am thinking of a large provision store in the Midlands, which also runs a restaurant. The owners about four years ago decided that it would be worth the risk to start a department for well-designed modern pottery, glass, rugs, and a variety of other articles suitable as gifts. They scarcely did this in order to obtain a sensational financial success, but rather to enable the public to acquire good modern things at reasonable prices. In fact, the result was that for the first two years a loss had to be booked—a loss covered by some of the profit of the other departments. Since then this department supports itself. A similar development has incidentally resulted in what is at present London's most fashionable shop for furnishing and decorative art. Such cases bring out what is one of my chief objectives in this book, namely that the success of any campaign for improvements in industrial art depends in the end on the goodwill of retailers and manufacturers.

As long as producer and distributor regard it as their sole object in business life to make money, it is a hopeless task to plead for better conditions of design. The manufacturer and the retailer who say that they are proud of not having any taste, that they are ready to supply anything which is asked for, and that educating the public is no business of theirs, must be called public enemies plainly and bluntly. They must change their outlook, or sooner or later public enterprise will interfere and make life rather unpleasant for them. It is not, I repeat, the object of this book to lessen the sales of existing firms; but what is required is that producers and distributors may begin to see the existence of two equal duties in their lives: one towards their private firms, their families, their incomes, and one towards the community.

It is a mistake that has been made time and again to lure manufacturers into improvements of design by promising them miraculous sales. The assertion of better design leading by itself to increased sales is to be found in discussions and manifestos everywhere since the eighteenth century, and it is prevalent still to-day. Unquestionably there are some cases where things have happened like that. I refer only to Orrefors and Ekco. But it would not be honest to pretend that these fabulous consequences of improved design are the rule and not rare exceptions. The truth is that the question should be put differently right from the outset. Manufacturers and retailers ought to endeavour to keep sales steady and yet to improve design. Are they willing to do this?

This question, a question of business morality, of commercial as well as of social pride, is at the bottom of the problem with which the present work has been dealing. Here is the only immediate remedy that can be recommended, while all other remedies, such as slum-clearance, better and more equal education, better and more practical art education, more intense propaganda, etc. are bound to act gradually and slowly. It rests mainly with the individual manufacturer and the individual retailer whether Industrial Art in England will remain as objectionable as it still is to-day, or whether it will improve, reach the best Continental standards, and perhaps even surpass them.

ADDENDA

The Finnish Bent Plywood Furniture (Better Design and Sales) (pp. 44, 210).

An exceedingly important and telling example, where the introduction of a new type of article has almost immediately led to increased sales, is that of the Finnish bent plywood furniture known in England as Finmar products. These were originated by Alvar Aalto, one of the most intelligent of Scandinavian architects. He began to experiment in 1928 assisted by an imaginative manufacturer, Mr Korhonen. Thus a technique was evolved which is based on the pressing into moulds of laminations of birch glued together, a technique highly suitable for inexpensive mass production. The first foreign exhibitions of Aalto's chairs, stools, etc. were held at Zurich and London in 1933. Since then demand has grown more rapidly than present supplies can cope with. In spite of the kind assistance of Mr N. G. Hahl, I have not been able to get exact figures to prove this. However, the wide popularity for modern flats and small houses of the Finnish birch stools selling at 8*s.* 6*d.*, or the arm chairs at 30*s.* and the occasional tables at 15*s.*, 21*s.* and 30*s.* is obvious enough. This, then, after the Orrefors glass, the tubular steel furniture, the Ekco wireless sets, would be a fourth case to prove the potential commercial value of new, good and consistent design.

Oriental and Modern Carpet Patterns (p. 62).

According to *The Furnishing Trades Organiser* (February 1937) the recently founded Institute of Carpet Manufacturers of America carried out an investigation on carpet design early in 1936. They sent round questionnaires to hundreds of distributors and published the following figures:

		Axminster	Wilton
		%	%
Oriental	1935	42·0	75·4
	1936	28·4	58·6
Chinese	1935	14·5	7·8
	1936	21·3	9·1
Colonial	1935	19·0	2·8
	1936	17·4	9·1
Textured	1935	—	—
	1936	8·6	8·0
Modernistic	1935	14·5	9·1
	1936	15·1	4·4
Floral and Leaf	1935	10·0	4·6
	1936	3·9	3·6

Art Education (pp. 152, 223).

In the new report of the Pick Council it is stated on p. 35 that the total number of students attending art schools in this country in 1935 was 31,851 men and 30,164 women. Only 7 per cent. of the men and 11·6 per cent. of the women were full-time students.

Another important document on art education published early in 1937 is the report of the Committee on Advanced Art Education in London appointed in 1936 by the Board of Education (Chairman The Viscount Hambleden). The foremost recommendation of the Committee is that "a new orientation should be given to the Royal College and that it should take the advanced study of all forms of applied art for its primary purpose" (p. 14). This should be done without giving up the teaching of the Fine Arts which cannot fail to exercise a broadening educational influence on the students of design. The London Central School should also essentially remain what it is. Only where the Central School has especially well-equipped classes (silversmithing, printing) should students of the Royal College attend courses at the Central School.

Further recommendations include the establishment of more design and craft classes, a closer co-operation between the departments of design and of architecture, the introduction of private studios for the teachers of the College, who should all be practising artists, and the arrangement of one-year courses for promising designers already active in industry, who should for this purpose be released by their employers and provided with scholarships.

The Committee wants to restrict the number of students enrolled at the College to 200 or 250. It pronounces itself in agreement with the Bauhaus system, i.e. a plant consisting of a comparatively small number of "basic" machines only, but close contacts with manufacturers who would admit advanced students to their works.

The existing courses for teachers should be reduced or even abolished.

National Register of Industrial Art Designers (p. 158).

A further proof of my assertion that the work of the Pick Council is not likely to remain a dead letter is afforded by the fact that, following a recommendation by the Council, a semi-official National Register of Industrial Art Designers has been set up in London, endowed with a Treasury grant. If this Register will be discriminating in the selection of those who are registered and put into touch with manufacturers, it should become a strong and beneficial power in the campaign for better design.

The Most Recent Activities of the Pick Council (pp. 158, 172, also p. 143).

On 20 January 1937 a report called *Design and Designer in Industry* was published, an admirable piece of writing, clear, succinct and comprehensive. It is based on the results of thirty-five meetings of the Council and on evidence received from seventy witnesses. On p. 7 of the report the suggestions contained in the report are summed up as follows: "The Report suggests, on the one hand, that manufacturers should help to secure better art education for young designers and for all ranks up to the heads of firms themselves, should employ in the design-rooms more fully-trained designers and give them better pay and prospects, and should secure the part-time services of skilled artist-craftsmen, and of selected architects, painters and sculptors, and on the other hand that all experts in art education should seek to understand the industrial point of view more fully, that the courses of training in our art schools should be modified and extended for industrial purposes and the teachers be secured a wider training and practical experience in industry, and that, by the closest co-operation between all concerned, the local art schools and colleges, or their industrial departments, should in future become, in effect, an organic part of industry working enthusiastically for industry, albeit remaining centres of liberal art education."

All these points are fully discussed in the sixty pages of the report, and many quotations could be given to show that the views adopted by the council are in accordance with those expressed in this book. Space does not allow me to quote more than four sentences, a haphazard assortment:

"Industry must face the question of releasing its younger workers for day-time instruction and must cease to rely on evening classes attended after the day's work is done" (p. 40).

"The development of art education in this country, particularly in the sphere of industrial art, has to some extent lagged behind that of technical training" (p. 29).

"It appears safe to state that the rates of pay (for designers) are low except in some of the wealthiest firms or companies, and are insufficient to attract the ablest artists or craftsmen" (p. 26).

"In many cases a scheme of visiting or consulting designers, employed for suitable periods—say for two or three months annually—would most conveniently meet actual requirements" (p. 19).

An interesting appendix of the report gives the recommendations in regard to a Regional College of Art at Manchester put forward by a special sub-committee on the training of designers for the cotton and rayon weaving and printing industries.

The publication of the report on furniture for working-class dwellings is planned for the month of April 1937, and it seems likely that this will also be full of interest. During the summer of 1937 the results of deliberations on Public Collections of Industrial Art in Britain and on the Continent will be made public.

Trade Papers and Design (p. 171).

The following is a quotation from *The Furnishing Trades Organiser* (February 1937): "The Furnishing Trades Organiser has one object. It wants to help the furnishing industries to sell more to the public....It...doesn't stand for good design in industry (as some apparently think). Nor does it stand for bad design (as some may have said)...."

The Furnishing Trades Notice and Welcome a Return to Period Forms (pp. 206–7, also p. 30).

The Cabinet Maker (26 December 1936) writes: "It has become increasingly apparent that as far as the best class of furniture is concerned designers have turned their minds from sensationalism towards beauty, and have tempered their advances with a respect for the work of their predecessors. They have drawn with discretion on typical motives of many styles, particularly those of the Empire and Regency periods, and have been wise enough to simplify and adapt rather than to copy slavishly. Sheraton in sycamore perhaps supplies a significant phrase to sum up the spirit of the serious designers' intentions during the year."

www.ingramcontent.com/pod-product-compliance
Lightning Source LLC
LaVergne TN
LVHW080042210926
848481LV00011B/3259